Handbook of Yarn Technology

Handbook of Yarn Technology

Girilal A V
Athira V
Deeksha Bhule

2026
Agri Horti Press
New Delhi-110 084

ISBN: 978-93-49690-08-0 (Hardbound)

Published By

Agri Horti Press

New Delhi

E-mail: agrihortipress@gmail.com

Digitally Printed at : Replika Press Pvt. Ltd.

Preface

Yarns form the fundamental building block of the textile industry, serving as the link between raw fibers and finished fabrics. As technology continues to evolve, so too does the science and engineering behind yarn production, processing, and application. A deep understanding of yarn technology is essential not only for textile engineers and manufacturers but also for researchers, students, and industry practitioners aiming to enhance quality, performance, and innovation in textile products.

This book, titled Handbook of Yarn Technology, is designed to provide a detailed and structured exploration of the key principles, techniques, and advancements in yarn manufacturing. It serves as a comprehensive reference for those involved in textile science, offering both foundational knowledge and current developments in the field.

Beginning with an overview of fiber types and their properties, the book systematically covers spinning processes including ring, rotor, air-jet, and friction spinning. It further delves into yarn structures, twist mechanics, yarn evenness, tensile properties, and fault analysis. Each chapter aims to blend theoretical explanations with industrial practices, giving readers a balanced understanding of both the science and application of yarn production.

Handbook of Yarn Technology also highlights emerging technologies and trends such as smart yarns, high-performance yarns, and sustainable processing techniques. The inclusion of modern testing methods, automation in spinning, and quality control standards ensures relevance to current industry needs and future innovation.

Practical insights, diagrams, and case examples enrich the content, helping readers to visualize and apply complex concepts. The book also pays attention to environmental considerations, waste management, and energy efficiency, reflecting the textile industry's growing commitment to sustainable production.

By bringing together diverse aspects of yarn technology in one volume, this handbook is intended to serve as both a learning tool and a professional reference. Whether used in academic settings or in the textile industry, Handbook of Yarn Technology aims to support knowledge development, technical excellence, and innovation.

We hope this book will be a valuable resource to all readers seeking to advance their understanding and expertise in the dynamic world of yarn technology.

Girilal A V | Athira V | Deeksha Bhule

Contents

Chapter 1

Review of Yarn Production

1.1 Introduction

1.1.1 Historical basis

It is very difficult to ascertain the historical period precisely by which man first started spinning fibres into yarns. However, based on the archaeological evidence, we can understand that this particular skill was well practiced at least 8000 years ago. Certainly, the weaving of spun yarns was developed around 6000 BC., the approximate time around which Neolithic man began to settle in permanent dwellings and to farm and domesticate animals.

We can safely assume and guess that early men would have twisted a few fibres from a lock of wool into short lengths of yarn and then tied them together to make longer lengths. We term these yarns as staple-spun yarns, because the fibres used are generally known to as staple fibres.

Possibly the yarn production would have been done by a set of two people working together as a team wherein one would be cleaning and spinning the wool, the other would be winding the yarn into a ball. As the various textile skills developed, the idea for spinning continuous knotless lengths would have led them to a stick being used, maybe first for winding up the yarn and then to twist and wind up longer lengths, thereby replacing the making of short lengths tied together and needing only one operator. This method of spinning a yarn using a dangling spindle or whorl was first developed as a handicraft and with the passage of time, it was widely practiced for processing both animal and plant fibres.

The simple spindle spinning technique continued as the only method of making yarns for a long time. Around AD1300, the first spinning wheel was invented and was developed in Europe which was termed as "the great wheel" or "one-thread wheel." The actual mechanization of yarn spinning took place over the period 1738 to 1825 to meet the huge rise in the demand for spun yarn resulting from the then-spectacular increase in weaving production rates with the invention of the flying shuttle . Pairs of rollers were introduced to reduce the fibre mass into a fine ribbon for twisting; spindles were grouped together on a frame to be operated by a single power source—the "water frame" , the "spinning jenny" and the "mule" followed by the "self-acting mule" by Roberts .

In 1830, a new method of inserting twist, known as cap spinning, was invented in the U.S. by Danforth. In the early 1960s, this was superseded by the ring and traveller, or ring spinning, which, despite other subsequent later inventions, has remained the main commercial method and is now an almost fully automated process of yarn spinning.

The long lessons of history show how the prosperity of a nation varies with the business activities of its population. The course of prosperity has always been bumpy and there are great dangers in extrapolating the future based on the short-term past. Successive centuries have seen fundamental changes of varying types in different parts of the world. Greenwood outlined the various steps or processes related to yarns and textiles in the first two millennia. It was pointed out that the extraordinary fine textile materials that have been made by craftsmen.

The mankind had always tried to develop and make machines to do the labourious and repetitive tasks of the textile manufacture by machines. Some of these developments have improved the productivity of the textile manufacturing systems and reduced the costs of textile production over the centuries. The eighteenth century witnessed the financial revolution followed by the industrial revolution in the nineteenth and the twentieth century paved the way for the information revolution.

The historical evidences of humanity have many references to textile materials because they were, and still are, part of the fabric of the lives of mankind. Consequently, the history of fibres is one of the easily traceable threads in the story of yarn manufacturing. The second thread of the history recorded the extraordinary developments that had taken place during the period of the industrial revolution. There were gigantic steps for the improvement of productivity of both people and machines especially in the textile industry. The list given below shows the details of the most important developments.

History has enough leads to make us think that the most important factor for success is to recognize expanding markets and plan accordingly. It was observed that the mechanical inventions had a relatively small effect on the overall competitive position. Other technologies, such as telecommunications and computing, are bound to have a greater effect. Nevertheless, the need to operate a mill in the most economic manner is still a very consideration, and high productivity machinery has to be used for major installations of the textile industry. In addition to this, there is a great need to produce products of consistent quality that will satisfy the requirements of the market. This is possible by involving quality control systems, which have become even more sophisticated. Ultimately the cost and quality of the product have to be carefully balanced for each market to achieve a competitive position without which the enterprise will fail.

Today, yarn spinning or production is a highly advanced technology that enables the engineering of different yarn structures having specifically desired properties suited for particular end use applications. The end uses include cover a broad range starting from garments for everyday use, household textiles, carpets, sports clothing, and fabrics for automotive interiors, aerospace, and medical and healthcare applications. A detailed understanding of how fibre properties and machine parameters are employed to produce yarn structures of appropriate properties is, therefore, an important objective in the study of yarn spinning technology.

1.2 Major developments of textile production

The major reason for the reduction in HOK over the years is that the productivity of the modern machines had increased considerably. Greenwood observed that the HOK values were ranging from 12 500 in Neolithic times, and were about 3120 in 'pre-fourteenth century'. It came to 0.63 for open-end (OE) spinning in the 1970s. In staple spinning, the invention of the mule was superseded by invention of the ring frame and then by the much newer technology. The technological shift to rotor spinning and other new technology in the USA and in some other areas had been highly significant. The productivity of the fastest machines had increased rapidly so far. As of now, but it is difficult to predict how the pace can continue in the future.

Already there are signs that the productivity curves of the textile industry are flattening and they seem to be approaching maximum values asymptotically. The second major development had taken place in the materials handling area. At the beginning of the twentieth century, the whole process of textile spinning consisted of a myriad of steps, with human intervention at each and every stage. Gradually the number of steps of the textile spinning had been reduced stage by stage and now automatic handling had become very much common. Automatic handling of the material processing comes in several forms. It ranges from the pneumatic transfer of fibre, to the use of robots to carry packages between machines of successive processing. It is understandable that these developments have also contributed to the drastic reduction in labour. At the beginning of this century, it appears that we have almost reached the irreducible minimum number of stages and automatic transfer of textile material. The end result is that there were no more big savings in labour cost.

1.2.1 World market for yarns

According to Thomas et al. , the market for spun yarns will always be dominated by cotton. At the time of writing this motes, the market share held by cotton is about two-thirds of the world market and this has been reasonably stable in recent years. Nevertheless, this is not to say that there were no changes in the market; on the contrary, shifts in consumer demands and preferences, total cost structures, and forceful geographic migrations of the industry across the globe had established themselves as powerful agencies for change.

Europe had suffered severe losses in production capabilities whereas the Asian output had soared by these losses. Though the American output had increased, the character of the same had also changed. There was consolidation amongst the companies that might be seen as evidence of the sorts of pressures that had influenced and affected Europe. However, Europe is still the reigning world leader in the smaller market of long-staple spinning. The production of cotton is still very strong in the USA and this is one of the reasons why the industry there has maintained stability even during the new millennium. Plants for the Production of polymers and man-made fibres and filaments had been established throughout the world. It is noteworthy to observe that the synthetic fibre and filament production using manmade polymers in Asia had made remarkable strides at the expense of Western and developed countries.

1.3 Introduction to Yarns

To know more about the yarn, we will look into the following:

1. The classification of yarns
2. Types of spinning processes and the spinning process variables
3. Different types of fancy yarns, their properties and end-uses
4. The relationships between the structures and properties of yarns

To understand and appreciate the process of yarn spinning, first of all we need know about what is a yarn. Therefore, we will try to know more about the product namely the yarn and them its manufacturing processes. The term yarn may be defined as a linear collection of filaments or fibres in a twisted state or tightly bound by other means, and possessing good mechanical properties. From the various types of commercially manufactured yarns, it can be seen that there are a great many functional and design possibilities of yarn production. Fibres are processed for the conversion to yarn in both pure and blended forms. Considerable variations in yarns that are made from a particular fibre or filament are possible. The classification of yarns (Table 1.1) is based both on their physical characteristics as well as performance properties.

1.4 Classification of Yarns

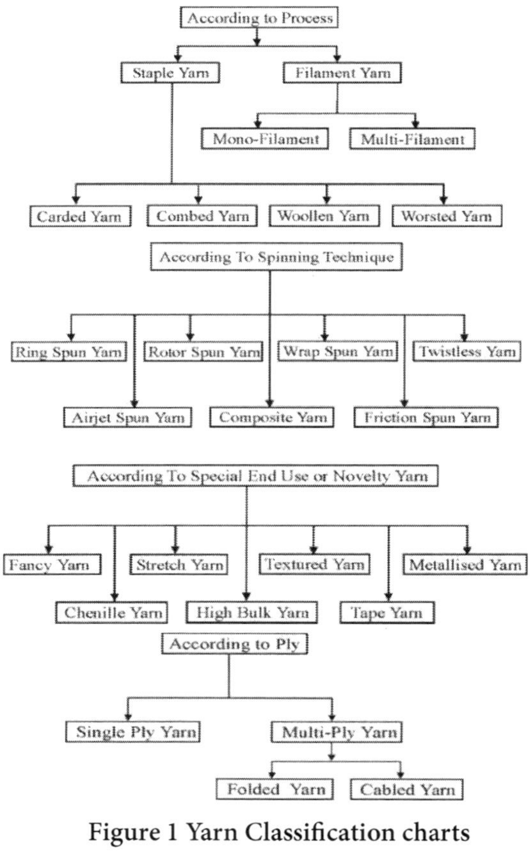

Figure 1 Yarn Classification charts

Yarns can be classified on various grounds such as a) nature of the component (fibre or filament) b) spinning method (ring spinning, open end spinning etc.) c) end use or novelty (fancy or surface effects) d) ply (number of constituent yarns). The relevant classification charts are given in Figure 1.

1.4.1 Staple Yarns

Staple yarns are defined as assembled strands of fibres twisted together to form a continuous strand of fibrous material of desirable properties. Currently, there are four types of staple fibre spun-yarn manufacturing systems commercially available. These are termed as carded, combed, woollen and worsted systems. Synthetic and regenerated fibres are stapled at the time

Table 1.1 Yarn types and their properties

Yarn type	General yarn properties
Staple yarns- Carded cotton, Combed cotton	Good hand, cover, comfort and textured appearance
Synthetic and blends	Average strength and uniformity
Worsted	
Woollen	
Continuous-filament yarns	High strength, uniformity and possibility for very fine yarns
Natural	Fair hand and poor covering power
Man-made or synthetic	
Novelty yarns	Decorative features and characteristics
Fancy	
Metallic	
Industrial yarns	Functional; designed and produced to satisfy a specific set of requirements
Tyre cord	
Rubber or elastic core	
Multiply coated	
High-bulk yarns	Great covering power with less weight, high loftiness or fullness
Staple	
Continuous filament (Taslan)	
Stretch yarns	
Twist-heat set-untwist	High stretchability and cling without high pressure, good handle and covering power
Crimp heat-set	
Stress under tension	
Knit-deknit	
Gear crimp	

Yarn type General yarn properties

Staple yarns- Carded cotton, Combed cotton Good hand, cover, comfort and textured appearance

Average strength and uniformity

Synthetic and blends

Worsted

Woollen

Continuous-filament yarns High strength, uniformity and possibility for very fine yarns

Fair hand and poor covering power

Natural

Man-made or synthetic

Novelty yarns Decorative features and characteristics

Fancy

Metallic

Industrial yarns Functional; designed and produced to satisfy a specific set of requirements

Tyre cord

Rubber or elastic core

Multiply coated

High-bulk yarns Great covering power with less weight,

high loftiness or fullness

Staple

Continuous filament (Taslan)

Stretch yarns of their manufacture and are then treated like natural fibres (cotton or wool) with different fibre lengths, diameters and crimps. This stapling process enables the mechanical processing to be carried out with few difficulties.

1.4.2 Continuous-Filament Yarns

Silk is the only natural continuous-filament yarn in use before the introduction of man-made fibres. Man-made filaments are manufactured by extruding the appropriate polymer solution through a spinneret, at which point the solution solidifies into a filament either by coagulation or cooling or evaporation. The number of orifices in the spinneret decides the number of filament in the bundle. The diameter and amount of drawing provided during the spinning crimped according to the required length for conversion to staple fibre, which will undergo further staple-spun yarn production.

CF yarns can be divided into untextured (i.e., flat) and textured yarns. CF textured yarns may be further grouped into several types; the more commonly used CF yarns are false-twist textured and air-jet textured yarns. For the first type, extruded filaments are stretched, then simultaneously heated, twisted, and untwisted, and subsequently cooled to give each filament constituting the yarn a crimped shape and thereby a greater volume or bulk to the yarn (see Figure 1.1).

Alternatively, groups of filaments forming the yarn can be fed at different speeds into a compressed-air stream (i.e., an air-jet), producing a profusion of entangled loops at the surface and along the yarn length. These processes are known as texturing or texturizing. The actual principle of false twisting texturization is used in other processes and is explained in a later section.

Continuous filaments can be cut into discrete lengths, comparable to the lengths of natural plant and animal fibres at the time of manufacture. Both manufactured fibres and natural fibres can be assembled (blended) and twisted together to form staple fibre-spun yarns. Table 1.1 shows that staple fibre spun yarns can be subdivided into plain and fancy yarns. In terms of both the Twist-heat set-untwist High stretchability and cling without high pressure, good handle and covering power Crimp heat-set Stress under tension Knit-deknit Gear crimp quantity used and technological importance, plain yarns are very popular in the industry. There is a wide range of differing types (i.e., structures) of plain yarns and thus the spinning techniques used to produce them. In the later chapters, we shall consider the production methods of both plain and fancy yarns.

1.4.3 Novelty Yarns

These yarns are specially designed and produced for specific decorative purposes and are seldom used to make an entire fabric, except in drapery applications. Most novelty yarns used are either of the fancy or metallic type.

Fancy yarns are in principle produced by the irregular plying of staple fibre or continuous filaments and are characterised by the presence of abrupt and periodic effects throughout the length of the yarn. The periodicity of these effects may be either irregular or constant. The desired novelty effect of the output yarn is brought about by a programmed difference in twist level or input rate in one or more components during the plying of the yarns. This results in differential bending or wrapping between the components of the yarns or in segments of buckled yarn that are permanently entangled in the resultant composite yarn structure.

1.4.4 Industrial Yarns

Industrial design requires special end-use yarns with specific functional characteristics. These yarns are engineered for performance under specified conditions. Many industrial yarns do not possess the visual and tactile properties of yarns used for apparel and home-furnishing applications. The best examples are tyre cord, asbestos and glass yarns, twine, rubber or elastic threads, core-spun yarns, wire yarn, sewing thread, heavy monofilaments and split-film yarns.

1.4.5 High-Bulk Yarns

A high-bulk yarn may be made up of either a staple fibre or continuous-filament yarn with normal extensibility but an unusually high level of loftiness or fullness. These yarns retain their bulkiness in both relaxed and stressed conditions. High covering power with lesser weight fabric is the characteristic feature of high-bulk yarns.

1.4.6 Stretch Yarns

Textured yarns which are pre-set for high extensibility at the time of production are termed as stretch yarns. While most of these yarns can be stretched up to twice their normal or relaxed length, some of these yarns can be stretched up to even three or four times their relaxed length. These yarns are both highly extensible and highly elastic. Most stretch yarns are produced by the texturizing of the thermoplastic continuous-filament yarns. This process results in reasonably good nonlinearity or crimp in the individual filaments of the output yarn. The resultant nonlinear structure of the filaments is heat-set but not entangled as in the case of the high-bulk yarns.

1.5 Staple-Fibre Yarns

A staple-spun yarn is a linear assembly of fibres, usually held together by the insertion of twist, to form a continuous fibrous strand that is smaller in cross-section but of a particular specified length. It is used for the manufacture of fabrics using the popular processes such as knitting, weaving and sewing. The actual strength or the quality of handle and appearance of both the yarn and fabric will depend on the way the fibres are assembled in the yarn system . The fibres used for spinning these yarns can either be natural or man-made. Manmade staple-fibre yarns are manufactured by using staple fibres cut from continuous filament at the time of synthetic fibre manufacturing before spinning. Staple-fibre yarns can further be subdivided and classified on the basis of staple fibre length, spinning method used for the manufacture and the type and nature of yarn construction. Each of these categories can be further subdivided.

Staple-spun yarn can even be classified as either short staple or long staple based on the staple fibre length of the fibres used in spinning. A staple fibre can have a staple length of about 10 to 500 mm. Short staple fibre by definition has a maximum length of 60 mm (cotton fibre is a short staple fibre with its staple length at about 25–45 mm). As per the conventional definition, long staple fibre has a staple length of more than 60 mm (wool fibre is a long staple fibre at about 60–150 mm).

1.5.1 Spinning Methods

Currently, different spinning types of methods/techniques are available for the manufacture yarns, including ring-spun, rotor-spun, twist-less, wrap-spun and core-spun yarns.

- **Ring-spun yarns:** This is the most widely used method of staple-fibre yarn production. The fibres are twisted helically around each other to impart strength to the yarn with the help of spindle, ring and traveller as the main components of twist insertion and winding.
- **Rotor-spun yarns:** These are similar to ring-spun yarns and are usually made from short staple fibres. Rotor is the most important twisting element in this spinning technique and is based on open-end spinning principle. Rotor spinning method produces a more regular and smoother, though weaker, yarn than ring spinning.
- **Twistless yarns:** The fibres are held together by some chemically based adhesives, not by the twist, and are often laid over a continuous filament core.
- **Wrap-spun yarns:** These yarns are made using the staple fibres wrapped around by another yarn, which is usually a continuous man-made filament yarn. Basically, these

yarns are made up of two components namely the wrap yarn (as the outer component) and the staple fibres as the inner component of the yarn. The yarns can be made from either short or long staple fibres which form the core and the filament yarn wraps them in a helical formation on a continuous basis for the entire length of the yarn.

- **Core-spun yarns**: Core-spun yarns have a central core (which is usually a filament yarn) that is helically wrapped with staple fibres, and are produced in a single operation at the time of spinning. It is basically made up of two components. For example, a typical core spun yarn can have a cotton sheath (outer component) for handle and comfort, with a filament (often polyester) core (the inner component) for added strength; or cotton over an elastomeric core. It is the converse of the wrap spun yarn with the components suitably interchanging the position with each other.

1.5.2 Operations in Staple-Fibre Spinning

There are a number of operations in the process of conversion of fibres into yarns which need to be done in a specified sequence.

- **Mixing**: The term 'mixing' refers to the bringing together of more than one variety of the same basic fibre for improving the output yarn characteristics and for optimisation of the mixing cost. For example, Egyptian cotton fibres can be mixed with American cotton fibre so as to get the optimum properties of both the fibres and the resultant yarn remains as ax100% cotton yarn.
- **Blending**: Blending refers to the bringing together of fibres of different types for the purpose of staple fibre spinning. For example, wool and silk or cotton and polyester fibres can be blended with each other.
- **Cleaning and fibre separation**: Bales of raw cotton fibres contain a variety of impurities, which must be removed in the early stages of spinning preparation so as to have good quality of spun yarns. The first process divides and splits the bales into loose bunches of fibres of reduced tuft size so as to remove dust, seeds and other unwanted contaminants. Some fibre types are then washed or scoured and others can be combed or carded to further separate and clean the fibres as per the requirements.
- **Fibre alignment:** This process follows carding and combing. Several slivers or groups of carded or combed fibres are combined together and then attenuated to form a single sliver of straightened/oriented fibres. This spinning preparatory process is known as drawing.
- **Drafting and twisting:** Drafting is the process of gently stretching out the slivers/roving to reduce their linear density or thickness to the desired level. The exact method and machinery used will depend upon the required yarn quality and count. In the final process of spinning, the required amount of twist is inserted into the single yarn which is wound on to the packages immediately on production.

1.5.3 Yarn Structure

The following are three major characteristics of a yarn:

1. It is a linear assembly of fibres. The assembly could be of any thickness
2. The fibres are held together usually by twist. However, some other means of inter fibre cohesion may be used.

3. There is a tendency for fibres of the yarn to lie in parallel along the twist spiral (helix). The most important aspect of yarn structure is its visual appearance, which is determined by the peripheral layer of fibres of the yarn. The second aspect of the yarn is its internal structure.

Yarn structures are generally very variable in nature. The differences are caused partly deliberately, according to the intended use of the yarn, but for the most part they are predetermined by the methods of yarn production available. The manner in which fibres are packed together in the yarn cross section is important to the effect of frictional contact between fibres on yarn properties. If fibres are loosely packed so that they can move about in the interstitial space, the yarn will appear bulkier and with a larger diameter than if fibres are closely packed. Two types of packing have therefore been assumed and proposed: Close packing is the packing which gives a hexagonal arrangement of the fibres in the yarn cross section, and open packing is the type of packing where the fibres are considered to be arranged in concentric circles of increasing radii. The basic helix model assumes an open packing configuration. For example, it is difficult to produce a yarn having properties similar to a ringspun yarn by new spinning processes, and ring-spun yarn still represents the standard reference for comparison with other types of yarns (Table 1.2) .

Yarn structure is primarily determined by the properties of the raw materials, the spinning process and parameters, the spinning unit conditions, machine parameters and settings and twist levels, and so on. The yarn structure can be open or closed; bulky or compact; smooth, rough or hairy; soft or hard; circular or flat; thin or thick, and so on.

In addition to appearance, yarn structure also influences the following yarn and hence fabric properties: handle; strength; elongation; insulating capacity; covering power; ability to resist wear, damage, strains, and so on; abrasion resistance; dyeability; tendency towards longitudinal bunching of fibres; and wearing comfort, and so on .

1.5.4 Applications of Staple-Spun Yarns

Table 1.2 Structures of Spun Yarns spun using different Spinning Methods

Yarn Type	Ring Spun		OE Spun		Air-jet Spun		Wrap Spun
Yarn sub class	Classic	Compact	Rotor	Friction	Two nozzle	One Nozzle	Filament wrapped
			Fibre Orientation				
Parallelism	Good	Very good	Medium	Low	Medium	Good	Very good
Compactness	Compact	Very compact: round	Open	Compact to open	Compact	Compact	Compact
Handle	Soft	Soft	Hard	Hard	Hard	Medium to hard	Soft
Hairiness	Noticeable	Low	Very low	Low	Some	Low to medium	Very low
Stiffness	Low	Low	High	High	High	Fairly high	Low
			Fibre Deposition				
In the core	Parallel. helical	Parallel. helical	Less parallel. helical	Less parallel. helical	Parallel w/o twist	Parallel w/o twist	Parallel w/o twist
In the sheath	Parallel. helical	Parallel. helical	More random. less twisted	Less parallel. helical	6% of fibres twisted around the core in spirals	20% of fibres twisted around core in spirals	Filament windings

Figure shows the count range for the different end-uses of staple-fibre spun yarns. In addition to the very fine yarn count range of 2–7.5 Tex meant for hosiery, staple-fibre yarns have almost similar market areas, where fine to medium yarn counts of 7.5–40 Tex are largely used to make textiles for clothing and apparel. Spun staple yarns hold a major position in the market for items such as shirts, blouses, home textiles, bed linen, trousers and suits .

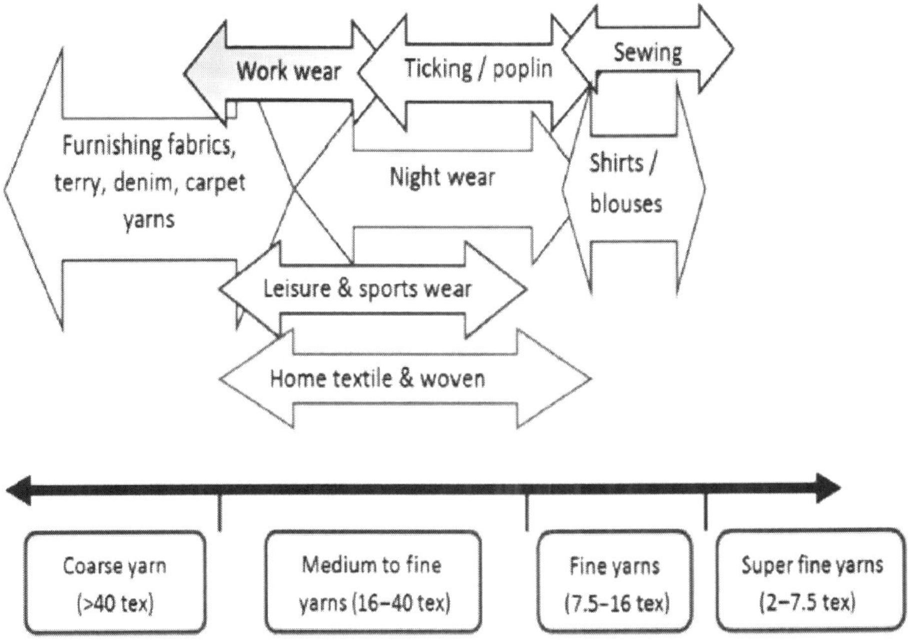

Figure 1.2 Chart showing the application range of different types of staple fibre spun yarns

1.6 Filament Yarns

A filament yarn is a collection of parallel bundles of filaments lying close together along the whole length of the yarn. Man-made yarns are made by extruding the required number of filaments in a single operation at the desired linear density. Yarns with only one filament are known as monofilaments and those with more than one filament are known as multifilaments.

1.6.1 Spinning Methods

Most synthetic fibres are extruded using polymers derived from by-products of petroleum and natural gas, which include polyethylene terephthalate (PET) and nylon as well as compounds such as acrylics, polyurethanes and polypropylene. The polymers are first converted from solid to fluid state by means of melting, dissolving using solvent, and so on. The fluid polymer is then extruded through a spinneret to convert the solution into filaments. There are four major types of synthetic fibre production techniques: Dry Spinning, Wet Spinning, melt spinning and gel spinning

The spinneret is a metal component having one to several hundred small holes. The fluid polymer is injected through these tiny openings to produce filaments from polymer solution.

This process of extrusion and solidification of innumerable filaments is known as the spinning of polymers. There are two types of extrusion: single-screw and twin-screw extrusion.

❖ **Single screw-extrusion**: This is one of the elementary tasks of polymer processing. The single screw extrusion process builds pressure on a polymer melt and forces it through a die or injects it into a mould. Most single-screw extrusion machines are plasticating in operation, taking solids in pellet or powder form and melting them while simultaneously building pressure.

❖ **Twin-screw extrusion:** This is commonly used for mixing, compounding or reacting polymeric materials. The flexibility of the twin-screw extrusion tool allows the operation to be designed specifically for the formulation being processed. For example, the two screws may be co-rotating or counter-rotating, inter-meshing or non-inter-meshing. The design and configuration of the screws themselves may also be changed using forward conveying elements, reverse conveying elements, kneading blocks, and other designs that can assist in obtaining particular mixing characteristics.

1.6.2 Polymer Spinning Processes

There are a number of spinning processes.

❖ **Wet spinning**: Of the four processes of chemical spinning, the oldest is wet spinning, as shown in Figure 1.3. This method is used for polymers that need to be dissolved in a solvent before they can be spun. The spinning solution is pumped though the spinneret. The spinneret is kept submerged in a coagulating chemical bath, which causes the polymer to precipitate at the time of spinning by dissolving its solvent and then solidify as it emerges through the spinneret holes. (The name of the spinning process is derived from the use of this 'wet' bath.) Acrylic fibre, Rayon fibre, Aramid fibre, Modacrylic fibre and Spandex fibres are all manufactured by the wet spinning process.

❖ **Dry spinning**: Dry spinning is used for polymers that need to be dissolved in a solvent. However, in this method solidification is obtained by the evaporation of the solvent. After the polymer is dissolved in a volatile solvent, the solution is pumped through a spinneret. As the spinning solution emerges through the spinneret, air or inert gas is used to evaporate the solvent from the solution to form the fibre by solidification. This resultant solidification of the fibres, which can then be collected and wound on a take-up wheel. The fibres are drawn to provide orientation to the macromolecular polymer chains along the fibre axis. This technique is used only for polymers that cannot be wet-spun due to the safety and environmental concerns associated with solvent handling. Dry spinning may be used for manufacturing acetate fibre, triacetate fibre, acrylic fibre, modacrylic fibre, PBI, Spandex fibre and vinyon. The process flow of dry spinning is shown in Figure 1.4.

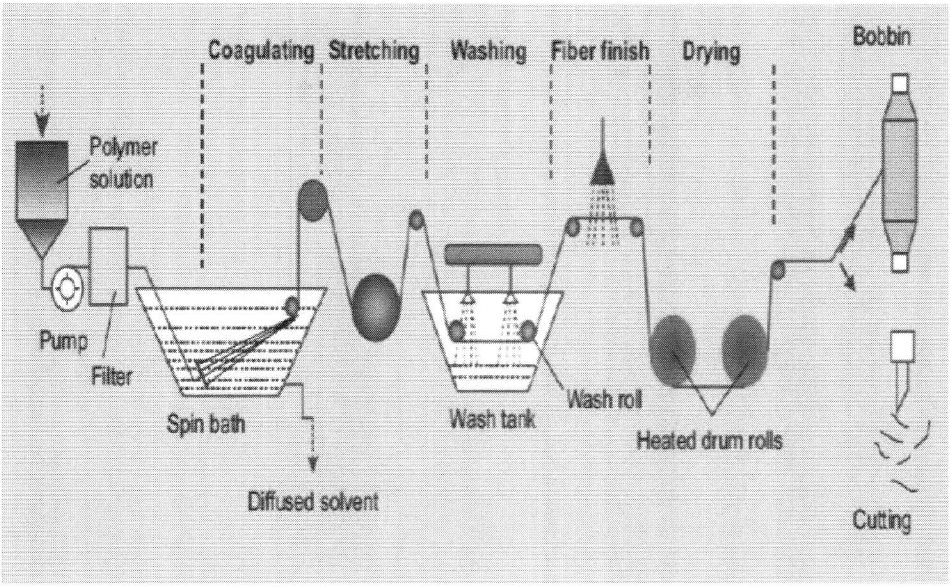

Figure 1.3 Schematic lay out of wet-spinning process

- **Melt spinning:** In this process, the thermo plastic fibre forming polymer of the fibre forming material is melted and then extruded through a spinneret. The molten fibrous material is cooled and solidified fibres are then collected on a take-up wheel. The fibres are stretched in the molten and solid states, which assists the orientation of the polymer chains along the fibre axis. Meltspun fibres can be extruded through a spinneret in different cross-sectional shapes, including circular, trilobal, pentagonal and octagonal as per the orifice design of the spinneret. The cross sectional or the physiological features of the fibres are responsible for some characteristic physical properties of the fibres. For example, trilobal-shaped fibres are capable of reflecting more light, which gives a lustre to the fabrics. Pentagonal-shaped and hollow fibres are soil and dirt resistant and are used in making carpets and rugs. Octagonal-shaped fibres offer glitterfree effects, while hollow fibres trap air, creating better insulation. The most popular fibre forming polymers such as polyethylene terephthalate and nylon 6-6 are melt-spun in high volumes. Nylon fibre, olefin fibre, polyester fibre, saran fibre, and so on are some of the thermoplastic synthetic fibres manufactured through melt spinning.

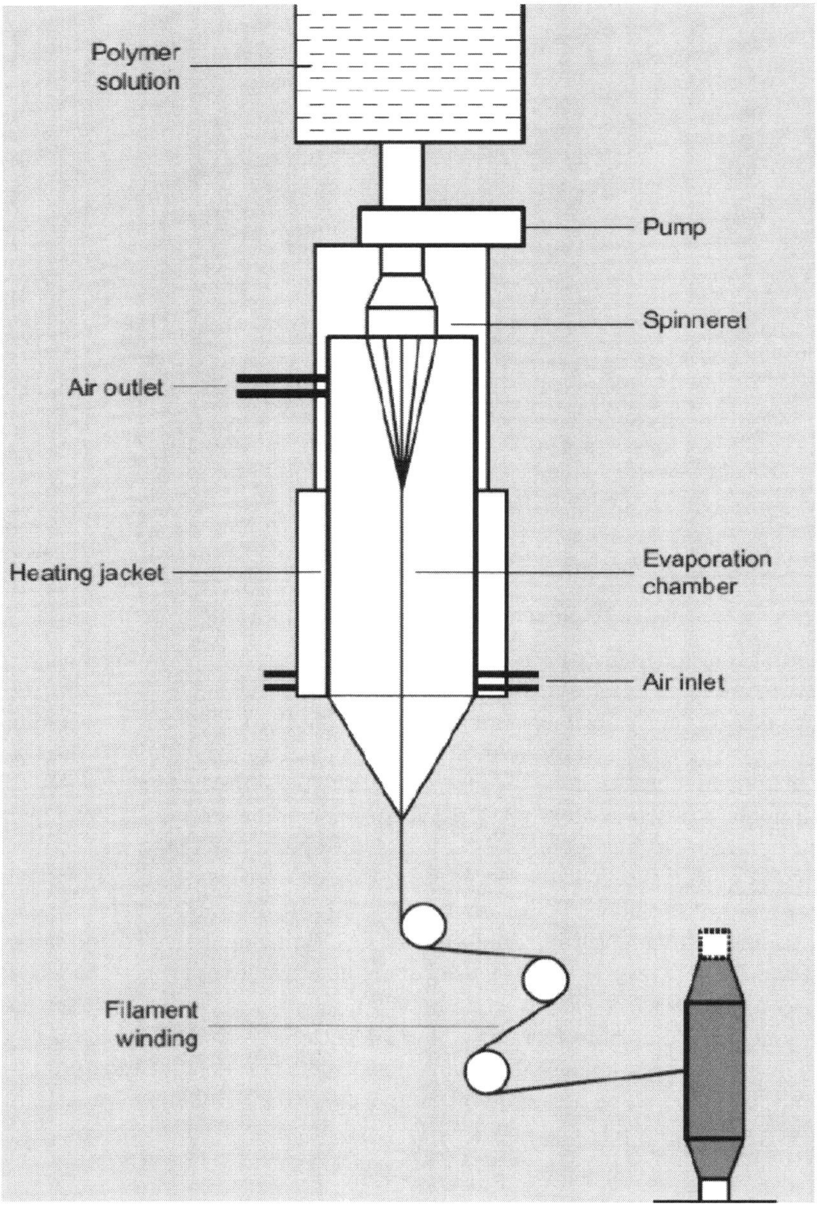

Figure 1.4 Schematic lay out of dry-spinning process

❖ **Gel spinning:** Gel spinning is also known as dry-jet-wet spinning, because the filaments first pass through dry air and are then cooled further in a liquid bath (wet). Gel spinning is the most widely used method to make very strong fibres with special characteristics. The fibre forming polymer is in a partially liquid or 'gel' state, which keeps the polymer molecular chains bound together to some extent at different points in a liquid crystal form. This bond results in strong inter-molecular forces within the fibre, which increases its tensile strength. The macromolecular polymer chains within the fibres also have a high degree of orientation, which further increases its strength. The strength is still further enhanced by the filaments emerging with an unusually high degree of orientation relative to each other. High-strength polyethylene and aramid fibres are currently manufactured by this process in the industry. The process flow of gel spinning is shown in Figure 1.5.

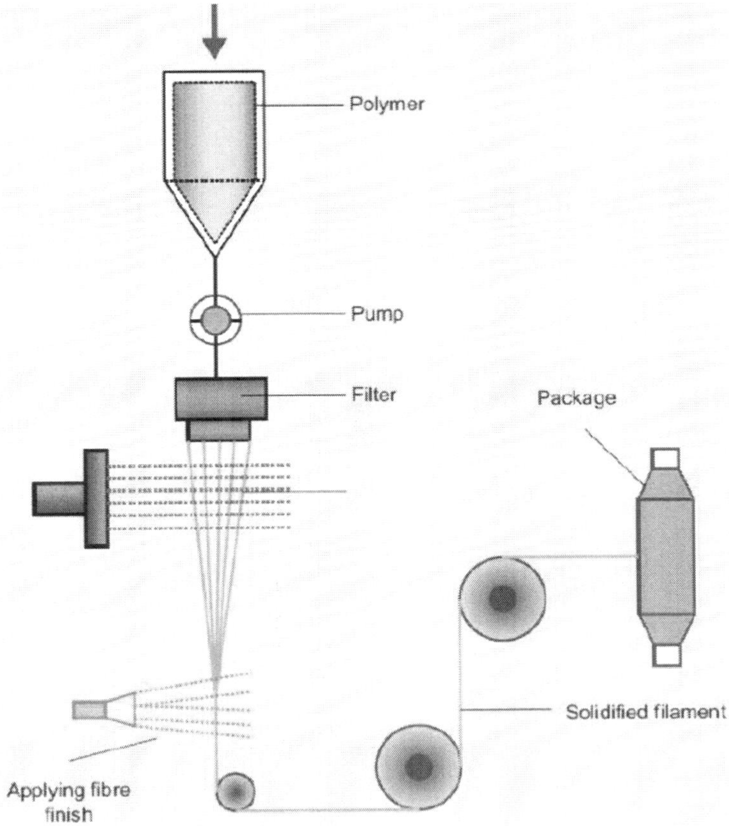

Figure 1.5 Schematic lay out of dry-spinning process

Whatever extrusion or spinning process is used for the chemical spinning of fibres, the fibres are finally drawn to increase both their strength and molecular orientation. This may be done either when the polymer is still in the process of solidification or after it has cooled completely.

Drawing forces the molecular chains to come together together and orients them along the fibre axis, resulting in a considerably stronger yarn. Man-made filament yarns can also be further divided into the following four subgroups depending on the physical constitution: flat, textured, bi-component and film (tape or split) yarns.

1.6.3 Structures of Continuous Filament Yarns

There are various types of continuous filament yarn structures currently developed and used by the industry:

- **Flat continuous filament yarns:** Standard (normal) filament yarns are known as 'flat' yarns, as compared to textured yarns, which may have the filaments in special curly or wavy form. They can be dull and matt or bright and lustrous as required.
- **Textured continuous filament yarns:** These are not to be confused with the either the fancy or decorative yarns. Textured filament yarns are man-made filament yarns in which the filaments have been specifically altered for a higher degree of bulkiness. Physical distortions such as crimps, loops and knots are introduced on to the filaments during the texturing process. Various texturing methods are used to introduce these deformations.
- **Bi-component continuous filament yarns:** These yarns are produced from two completely different fibre forming polymer components that are brought together at the fibre extrusion stage. Each component can be designed to have individual characteristics, such as differential shrinkage ratios, which can cause kinking or spiralling to imitate natural wool.
- Tape or split-film yarns: To make film, the polymer is extruded in a thin, wide sheet. The film is cut into narrow strips or ribbons to produce tape. Slits can then be made along the length of the ribbon to produce a split-film yarn.

1.6.4 Applications of Filament Yarn

Figure 1.6 illustrates the wide count range for the different end-uses of filament yarns. These yarns are highly competitive in the carpet-yarn and sportswear applications and in industrial yarn applications for technical textiles.

1.7 Fancy Yarns

1.7.1 Marl Yarn

The simplest among the fancy yarns, is the marl yarn and is made by twisting two differentcoloured yarns together in a yarn doubling process. It has a different texture from normal double yarn. These yarns are used to make discreet pinstripes in men's suiting or to produce a subtly and irregularly patterned knitted fabric with a relatively simple fabric construction. They may also be used to provide a Lurex or other metallic yarn with strong support, while at the same time creating a more subtle effect.

1.7.2 Spiral or Corkscrew Yarn

A spiral or corkscrew yarn is a plied yarn displaying the characteristic smooth spiralling of one yarn component around the other and is very similar to the structure of a marl yarn. Figure 1.10 shows the basic structure, which is straightforward, except in the differing lengths of the two yarns used.

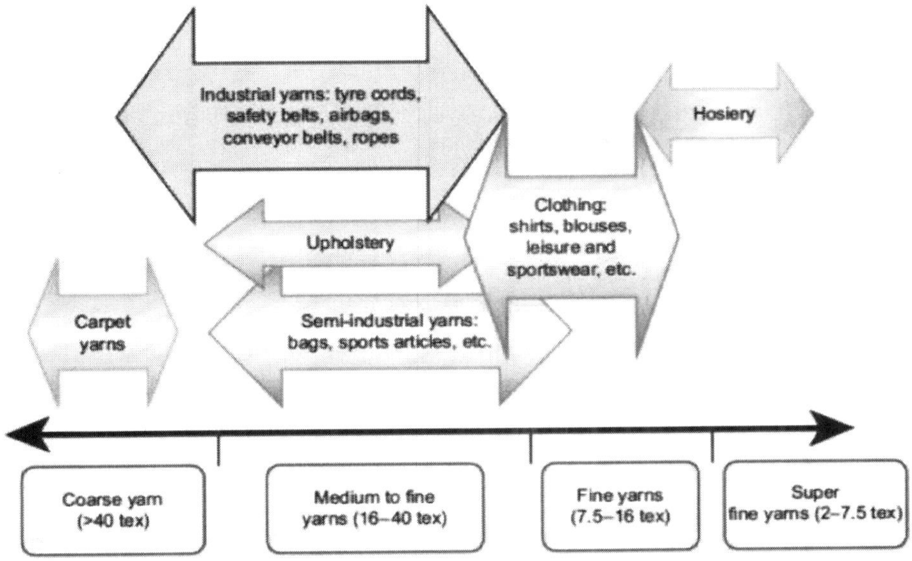

Figure 1.6 End-uses of filament yarns.

1.7.3 Gimp Yarn

A gimp yarn is a compound yarn consisting of a twisted core with an effect yarn wrapped around it so as to produce wavy projections on the yarn surface. This structure is shown in Figure 1.6. Because of the binder yarn requirement to give stability to the structure, the yarn has to be produced in two stages. Two yarns of widely varying counts are plied together, thick around thin, and are then reverse bound. The process of reverse binding removes the twist that creates the wavy profiles as it makes the effect yarns longer than the actual length of the completed yarn. The texture properties of a gimp are clearly better than that of a spiral yarn. The finer of the two gimps shows that the effect is less regular and perhaps less well-defined.

1.7.4 Diamond Yarn

A diamond yarn is produced by folding a coarse single yarn or roving with a fine yarn or filament of contrasting colour using S-twist. This is cabled with a similar fine yarn using Z21 twist. Multifold 'cabled' yarns may be made by extending and varying this technique to produce a wide range of effects. A true diamond yarn would show some compression effect upon the thick yarn from the thin ones, but in the interests of clarity this is not shown in Figure 1.7.Diamond yarn is very useful to designers in the creation of subtle effects of colour and texture, particularly in relatively simple fabric structures.

1.7.5 Boucle Yarn

This type of yarn is characterised by tight loops projecting from the body of the yarn at fairly regular intervals, as shown in Figure 1.10. Some of these yarns are made by air-jet texturing but most are of three-ply construction. The three components of the yarn are the core, the effect and the tie, or binder yarns. The effect yarn is wrapped in loops around a core or base yarn, and then the third ply, or binder, is wrapped over the effect ply in order to hold the loops in place. The individual plies may be filament or spun yarns. The characteristics of these component yarns determine the ultimate design effect.

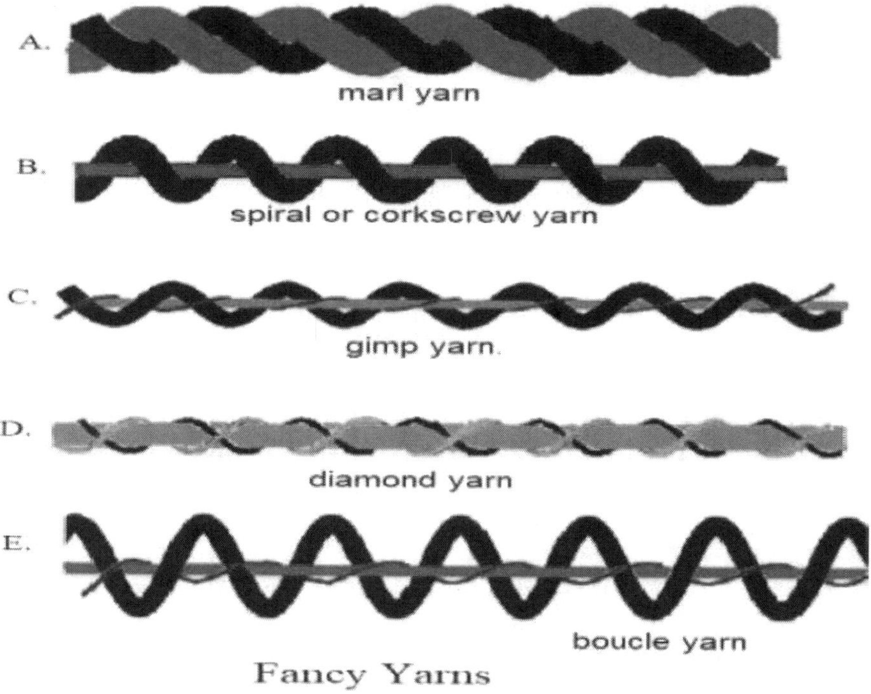

Figure 1.7 Different types of fancy yarns

1.7.6 Loop Yarn

A loop yarn consists of a core with an effect yarn wrapped around it and overfed to produce a nearly circular loopy projection on its surface. Figure 1.8 shows the structure of a loop yarn, in this case somewhat simplified by showing the core as two straight bars. In reality, the core always consists of two yarns twisted together, which entraps the effect yarn. As a general rule, four yarns are involved in the construction. Two of these form the core or ground yarns. The effect yarn(s) are formed with an overfeed of about 200% or more. It is important for these to be of the correct type and of good quality. Even, low-twist, elastic and pliable yarn is needed. The effect yarn is not completely entrapped by the ground threads and therefore a binder is necessary. The size of the loops is determined by the level of overfeed, the groove space on the drafting rollers, the spinning tension or the twist level of the effect yarn. Loop yarns can also be made with slivers in place of yarns for effect.

1.7.7 Snarl Yarn

Snarl yarn has a similar twisted core-to-loop structure. Again for the sake of simplicity, the core is shown in Figure 1.8 as two parallel bars. A snarl yarn displays 'snarls' or 'twists' projecting from the core. It is produced by similar method to the loop yarn, but uses a lively, high-twist yarn and a somewhat greater degree of overfeed as the effect yarn. The required size and frequency of the snarls may be obtained by careful control of the details of overfeed and spinning tension, and by the level of twist in the effect yarn.

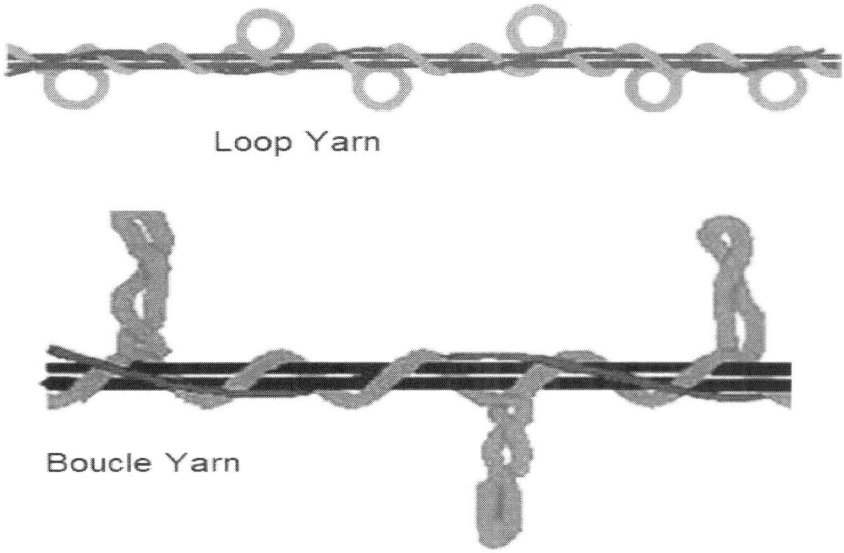

Figure 1.8 Loop yarn and snarl yarn

1.7.8 Knop Yarn

A knop yarn contains prominent bunches of one or more of its component threads, which are arranged at regular or irregular intervals along its length (Figure 1.9). It is normally produced by using an apparatus with two pairs of rollers, each capable of being operated independently. This makes it possible to deliver the base threads intermittently, while the knopping threads that create the effect are delivered continuously. The knopping threads join the foundation threads below the knopping bars. The insertion of twist collects the knopping threads into a bunch or knop. The vertical movement of the knopping threads forms a bunch or knop. The vertical movement of the knopping bars determines whether the knop is small and compact or spread out along the length of the yarn.

1.7.9 Slub Yarn

This is a yarn in which slubs are deliberately created to make the desired effect of discontinuity. Slubs are thick places in the yarn that may take the form of a very gradual change, with only a slight thickening of the yarn at its thickest point. Alternatively, a slub may be three or four times the thickness of the base yarn and the increase in thickness may be achieved within a short length of yarn. The yarn pictures in Figure 1.9 should give a clear impression of the structure of the yarn itself.

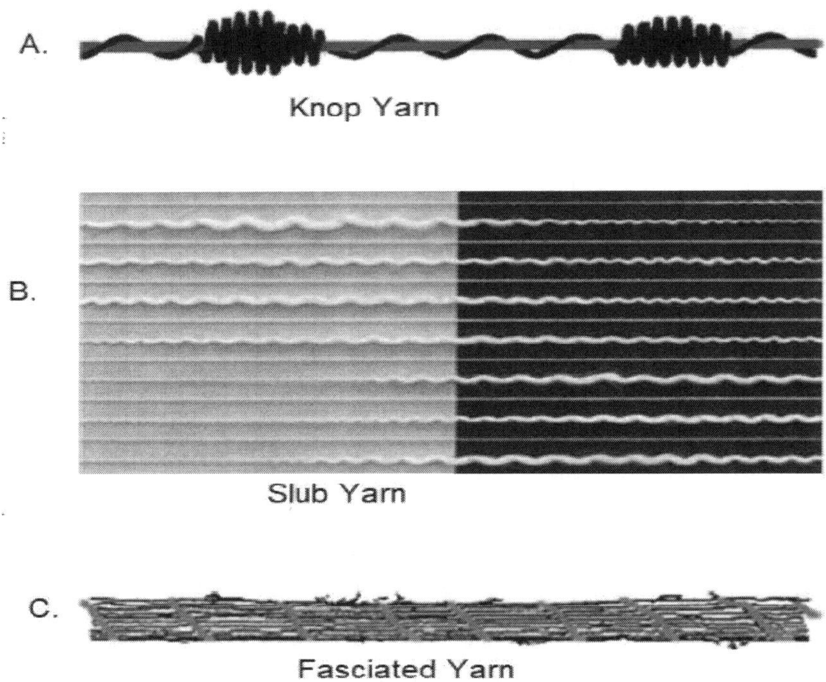

Figure 1.9 Knop, slub and fasciated yarns

1.7.10 Fasciated Yarn

This is a staple-fibre yarn that consists of a core of parallel fibres bound together by wrapper fibres. Yarns made by the air-jet spinning method are structured in this way. Yarns produced by the hollow spindle method are also frequently described as fasciated, as the binder is applied to an essentially twistless core of parallel fibres. It is possible to see fibres that have escaped the dark binding thread and contrast with one of the two slivers used as feedstock in making the yarn.

1.7.11 Tape Yarn

Tape yarns may be produced using various processes including braiding, warp knitting and weft knitting. In recent years, these materials have become better known, especially in fashion knitwear. It is also possible to use narrow woven ribbons, narrow tapes of nonwoven material, or slit film in the same way.

1.7.12 Chainette Yarn

Chainette yarn, shown in Figure 1.10, is produced in a miniature circular weft knitting process, often using a filament yarn and a ring of between 6 and 20 needles. The process has been used on a small scale for many years and is now used extensively in fashion knitwear.

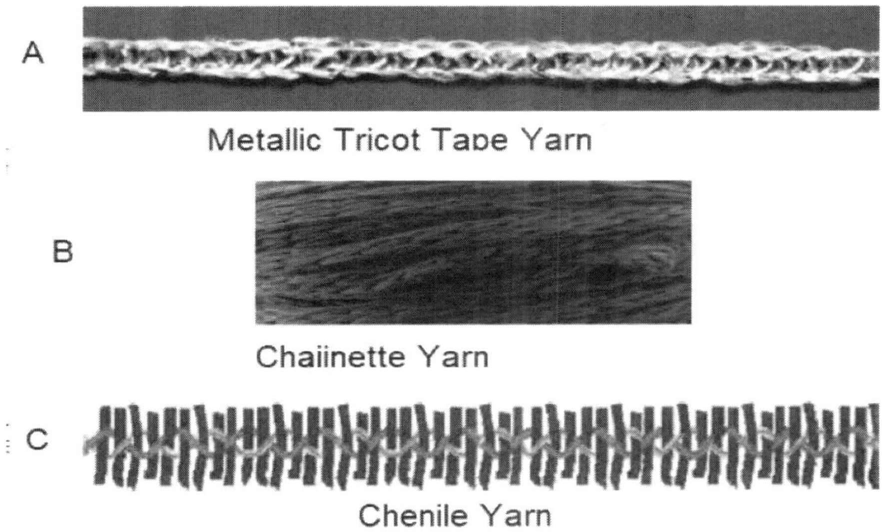

Figure 1.10 Tape yarn, chainette yarn and chenille yarn

1.7.13 Chenille Yarn

True chenille yarns are produced from a woven leno fabric structure that is slit into narrow, warp-wise strips to serve as yarn. These are pile yarns, and the pile length may be uniform throughout the length of the yarn or may vary in length to produce a yarn of irregular dimensions. Chenille yarns are used in furnishings and apparel. Chenille yarns, as shown in Figure 1.10, have a soft, fuzzy cut pile that is bound to a core. These yarns can be spun, but the machinery required is very much specialised. For this reason, these yarns are usually woven on a loom. The effect yarn forms the warp, which is bound by a weft thread. The weft thread is spaced out at a distance of twice the required length of pile. The warp is then cut halfway between each weft thread.

1.7.14 Ribbon Yarns

These yarns are not produced by spinning and consist of finely knitted tubes, pressed flat to resemble ribbon or tape. The ribbons are usually soft, shiny and silky.

1.7.15 Composite Yarns

Also known as compound yarns, these consist of at least two threads. One forms the core of the composite yarn, and the other strand forms the sheath component. One thread is a staplefibre yarn and other a filament yarn. Compound yarns are even in diameter, smooth and available in the same count range as spun and filament yarns.

1.7.16 Covered Yarns

Covered yarns have a core that is completely covered by fibre or another yarn. Figure 1.11 shows different types of covered yarns. The core might be an elastomeric yarn, such as rubber or Spandex, or other yarns, such as polyester or nylon. Covered yarns may have either a single or double covering. The second covering is usually twisted in the opposite direction to the first. Single-covered yarns have a single yarn wrapped around them. They are lighter, more resilient and more economical than double-covered yarns and can be used in satin, batiste, broadcloth and suiting as well as for lightweight foundation garments. Most ordinary elastic yarns are double-covered to give them balance and better coverage. Fabrics made with these yarns are heavier.

1.7.17 Metallic Yarns

These have been used for thousands of years. Metallic yarns may be made of monofilament fibres or ply yarns. Two processes are commonly used to produce metallic yarns. The laminating process seals a layer of aluminium between two layers of acetate or polyester film, which is then cut into strips for yarns, as shown in Figure 1.11. The film may be transparent, so the aluminium foil shows through, or the film and/or the adhesive may be coloured before the laminating process. The metallising process vaporises the aluminium at high pressure and deposits it on the polyester film.

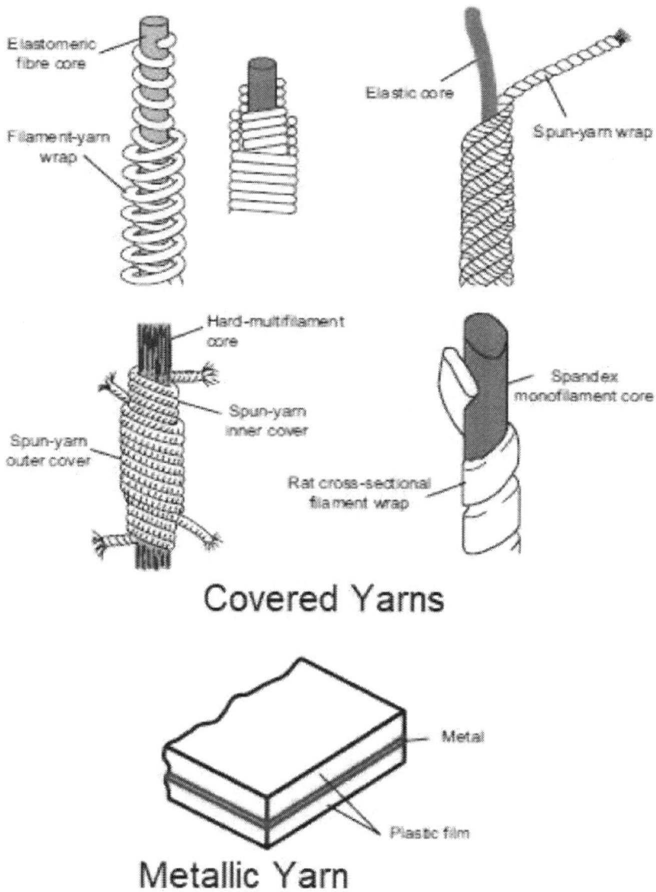

Figure 1.11 Covered yarn and metallic yarn

1.8 Staple-Fibre Yarn Manufacturing

An overview of different types of spinning methods is presented here and each method is given in greater detail in the Chapter 2.

1.8.1 Ring (Conventional) Spinning

Ring spinning is a process that converts the short, raw fibre into a continuous yarn using a series of machines, as shown in Figure 1.12. The conventional systems for converting staple fibre into spun yarns are those developed for cotton and wool: opening, carding, drawing, combing, roving and spinning.

- **Opening:** This is the first operation in the spinning of yarn from raw fibres. Opening is the basic process of reducing the tuft size of the compressed cotton fibres from a bale into smallerfibre tufts. It removes the particles of dirt, dust and other impurities from the cotton tufts by using spiked rollers. After this process the fibre will be transferred to another process.
- **Carding:** After mixing (blending) and opening, loose fibres are transferred to a carding machine. Carding is performed by opposing sets of teeth or small wire hooks known as card clothing, which cover the machine parts and include a licker-in, a cylinder, revolving flats and a doffer. The cylinder and the flats may rotate in the same or opposing directions but at different speeds to tease the fibre tufts into a thin, filmy web, which is then collected into a loose ropelike structure called a sliver, which is often coiled, and deposited in cans. Carding further opens the fibre tufts and extracts any fine particles, neps and short fibres enclosed by the fibre aggregates. The drawing frame uses a series of rollers arranged in pairs and rotating at different speeds. Slivers are passed between the rollers and combined. The fibres will be well parallelised and mixed after going through this process.
- **Combing**: Combing is the process used to remove short fibres and neps from sheets of cotton fibres (lap). A roller with fine-toothed elements fixed on a half-lap is used. The amount and length of the short fibres extracted will depend on the combing parameters selected. The fibres will be straightened and paralleled during this process.
- **Roving:** In this process, slivers are reduced to around one-eighth of their original diameter by three pairs of rollers, rotating at different speeds. The required level of twist is also imparted to keep the rovings stable under the stretching caused by winding and unwinding.
- **Ring spinning:** The conversion of roving in to yarn is called the spinning process. This is usually done in a roller drafting system that will have some means of fibre control, such as a double apron. Twist is imparted to the fibre strands to prevent slippage through the ring and traveller. The yarn is then wound onto suitable bobbins known as ring cops for further processing.

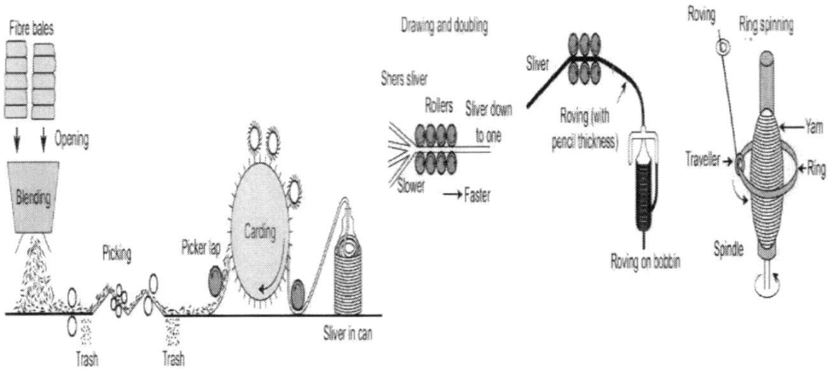

Figure 1.12 Process flow chart of conventional ring-spinning process.

1.8.2 Hollow-Spindle Spinning

In this process, the twist in a yarn is replaced by wrapping a filament binder around the staple fibre materials used (Figure 1.13). This results in a fasciated yarn structure in which most of the core fibres/yarns run parallel to one another along the axis of the strand, while the filament binder imparts the necessary cohesion. Despite superficial similarities, yarns made using the hollow-spindle system are quite different in structure from those made by the conventional ring-spinning system. They are also likely to differ in details of appearance and behavior during processing. Hollow-spindle yarns are used mainly in knitted garments or fabrics, although plain yarns have found many other applications such as carpets and medical textiles.

When used in the production of fancy yarns, the hollow-spindle technique adds the binder and immediately the effect is produced instead of using a separate second operation. In yarns produced on hollow spindles, there is no twist holding the core fibres/yarns of the fancy yarn together, so there is no cohesion beyond that imparted by the binder. If the binder breaks, the core fibres fall apart more freely and dramatically than would be the case with a fancy yarn produced by the ring-spinning system. In this particular example, there are four independent feeding devices, three for effect fibres and one for the core yarn. The effect fibres are fed in the form of staple roving or slivers.

The fibres are then drafted using a roller drafting system similar to that used on ring frames. The effect fibres are combined with the core yarns and then passed through the rotating hollow spindle. A bobbin bearing the binder, usually a filament yarn, is mounted on the hollow spindle and rotates with it. The binder yarn is pulled into the hollow spindle from the top.

The rotation of the hollow spindle wraps the binder around the staple strand and the core yarns. The binder then holds the effect and core yarns in place. To avoid the possibility of the drafted staple strand disintegrating before it is wrapped by the binder, the spindle usually generates a false twist in the staple strand. The staple strand does not therefore pass directly through the hollow spindle but is first wrapped around a twist regulator, which is usually placed at the bottom of the spindle.

A very wide range of fancy effects can be produced with the hollow-spindle system. Many of these effects can be controlled by controlling the speeds of the corresponding feeding devices. It is also possible to use the hollow-spindle system to create fancy yarns that include yarns in their effect. Many more effects can be produced by controlling the final yarn delivery speed.

Since the effect fibres do not have a real twist, hollow-spindle yarns differ from ring yarns in both their appearance and their performance characteristics. The former tend to be bulkier and have lower wear resistance

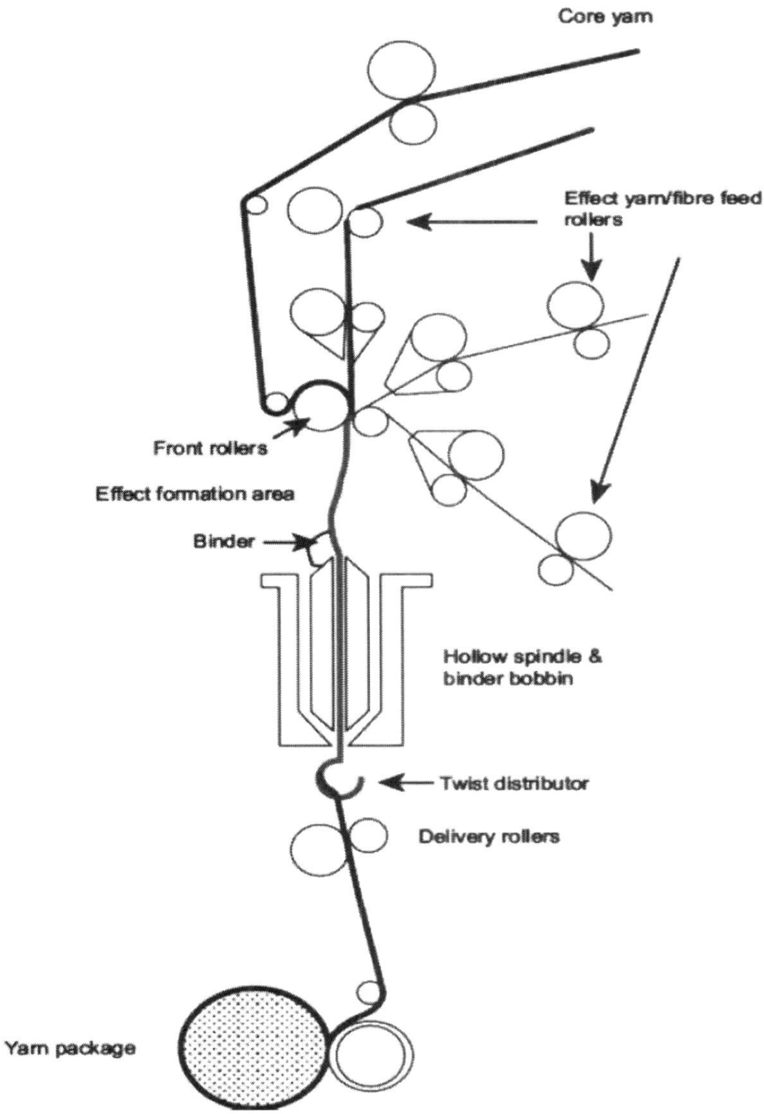

Figure 1.17 Schematic diagram of hollow-spindle spinning method

1.8.3 Combined Systems

Combined systems were first established in order to unite the benefits of the ring and hollowspindle systems in a single machine, as it was thought that a yarn with twist had a more stable and reliable structure than one with a fasciated structure. Later, it was recognised that two hollow spindles could also be assembled in series and that this would offer a variety of yarns and a different range of benefits. This is illustrated in Figure 1.14, which depicts two hollow spindles, arranged one above the other, which wrap the staple strand with two binders applied in opposing directions. This technique is used to produce special-effect yarns that have a more stable structure, as the effect fibres are trapped by two binders instead of one. In this case, the wrapped yarn is provided with some true twist by the ring spindle placed immediately below the hollow spindle. It was thought that the speed of the hollow spindle assembly would be enhanced by the true twist inserted by the ring spindle, and that it would therefore be able to create yarns that are less expensive than true ring-spun yarns while still retaining some of their characteristics.

1.8.4 The Doubling System

The conventional doubling system is based on ring spinning. The arrangement provides two or more yarns that can be fed independently at controlled speeds. These may include uniform, fluctuating or intermittent feeds as required, so permitting a simple means of producing spiral or marl-type yarns, although obviously requiring the feed material to be in yarn form. This method allows spinners who do not specialise in fancy yarn production to manufacture some of the simpler fancy yarn structures. The doubling frame can produce some interesting effects, particularly when it is used to combine two existing fancy yarns to produce another. It is also possible to manufacture spiral effects using an ordinary doubling system. This can be done by combining two yarns of very different counts and opposite twist. If the doubling twist is in the same direction as that of the thicker single-yarn twist, the thicker yarn contracts while the thinner yarn expands, thus causing the thinner yarn to spiral around the thicker one. If the doubling twist is in the opposite direction to that of the thicker single-yarn twist, the thicker yarn expands while the thinner yarn contracts, causing the thicker yarn to spiral around the thinner one. Although their basic structure is identical, these yarns are aesthetically very different and will be used in different ways.

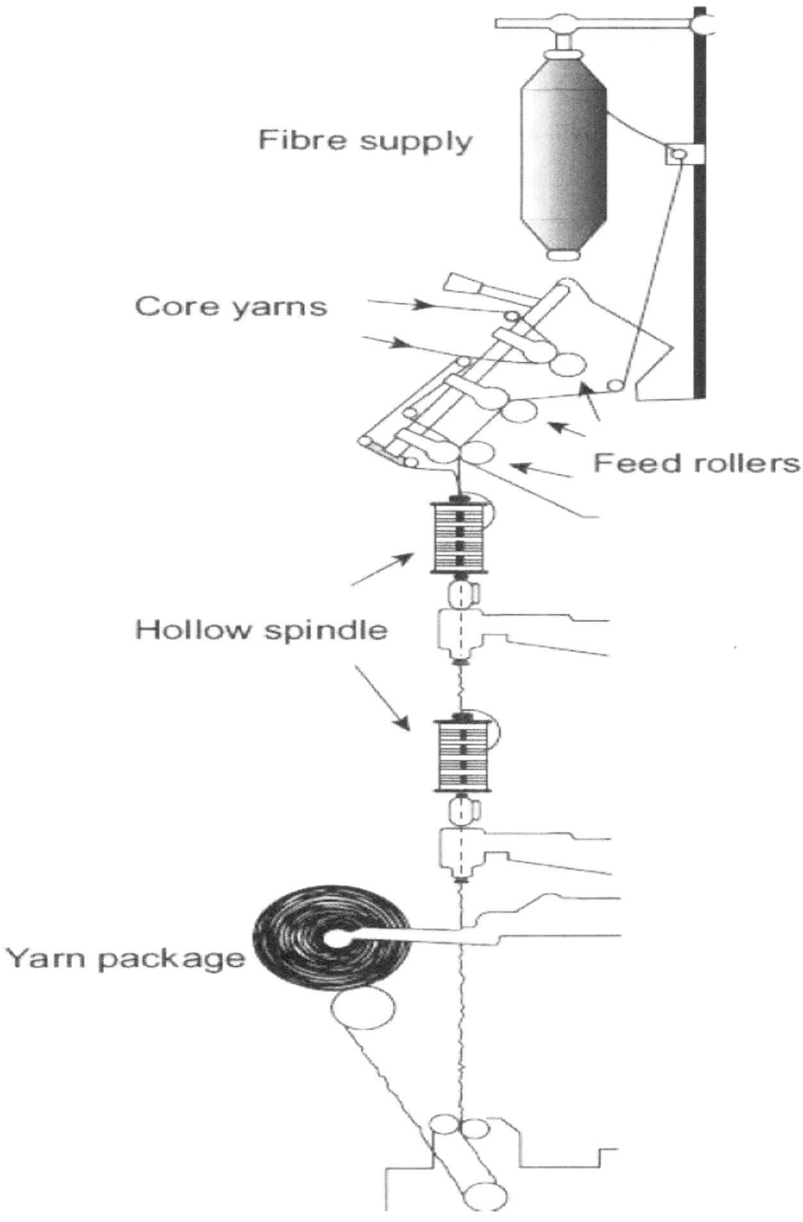

Figure 1.14 Schematic layout of two-spindle wrap system.

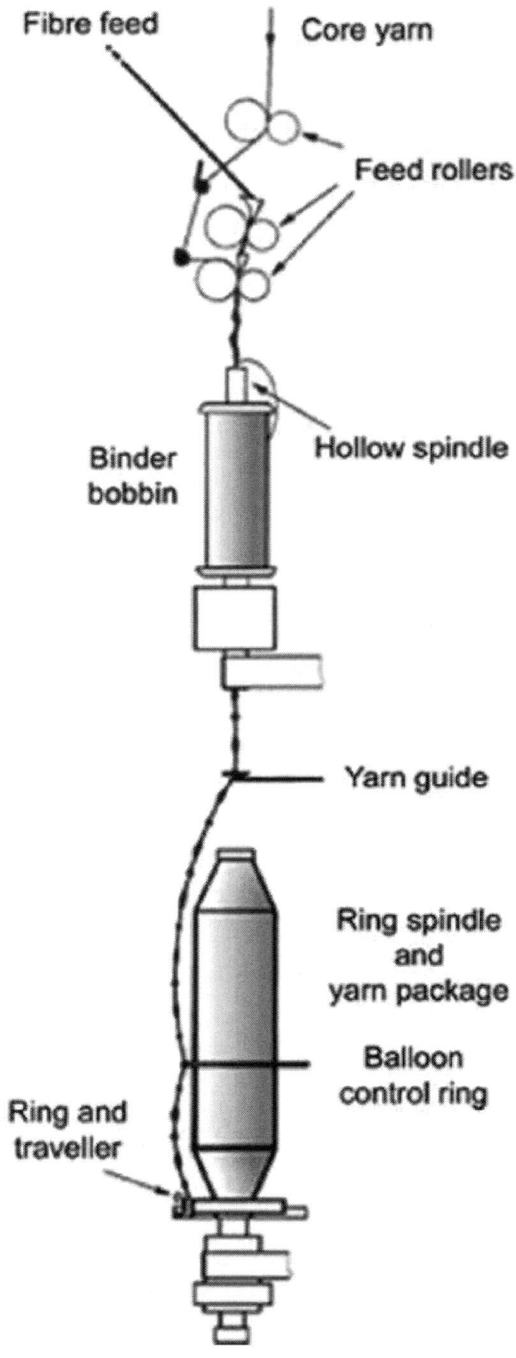

Figure 1.15 Combination of Ring and Hollow spindle

1.8.5 Open-End Spinning

There are two commonly used yarn manufacturing methods operating on the open-end spinning principle: rotor and friction. The rotor system is mainly used for the production of coarse to medium count short staple yarns. The friction system is used mainly to make coarser industrial yarns. However, both systems may also be used for making some fancy yarns.

1.8.5.1 Rotor system

In open-end spinning, the yarn twisting action is separated from the winding action and the package needs to rotate only at a relatively low winding speed. The process may be divided into the following steps: opening, transport, alignment, overlapping and twist insertion. In the rotor-spinning process, as shown in Figure 1.16, individual fibres are carried into the rotor on an air stream and laid in contact with the collecting surface so that a strand of fibres is assembled around the circumference. As the fibres are drawn off, twist is imparted by the rotor, to produce a yarn. Rotor spinning is most suitable for spinning short staple-fibre yarns. Recent developments in electronic control have allowed the development of rotor spinning machinery that is also capable of producing slub yarns. These yarns are used in furnishings and drapes, rather than in apparel fabrics, although they are sometimes used in denim fabrics. They are manufactured using attachments to ordinary open-end spinning devices, which usually incorporate an electronically controlled device for the brief acceleration of the drawing-in roller. As a result of the back doubling action inside the rotor, it is not possible to produce slubs shorter than the circumference length of the rotor because any variation in the fibre feed material is spread over a minimum length of the rotor circumference.

There have also been attempts to alter the fibre flow and thus introduce variations in the yarn appearance by the use of injecting pressurised air into the fibre-transportation tube. However, the effects created using this approach are very limited, as the fibre flow in the transportation tube is slight and the variation in the yarn caused by changes in the airflow is therefore very small.

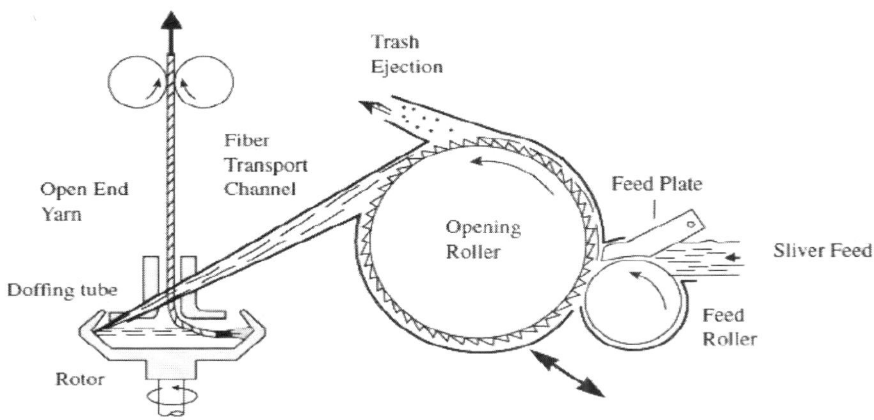

Figure 1.16 Schematic diagram on the Principle of rotor spinning.

1.8.5.2 Friction system

Friction spinning is a different open-end spinning technique when compared to rotor spinning. Instead of using a rotor, two friction rollers are used to collect the opened-up fibres and to twist them into yarn. The principle of DREF-2 is shown in Figure 1.17. The company also produces the DREF 3 machine, which has an extra drafting unit in the machine in order to feed drafted staple fibres to form a core component. The fibres are fed in sliver form and are opened by the opening roller. The opened fibres are then blown off the opening roller by an air current and transported to the nip area of two perforated friction drums. The fibres are drawn onto the surfaces of the friction drums by air suction. The two friction drums rotate in the same direction, and twist is imparted to the fibre strands because of the friction with the two drum surfaces. The yarn is withdrawn in the direction parallel to the axis of the friction drums and is delivered to a package-forming unit. A high twisting speed can be obtained even while using a relatively low speed for the friction drums, because the friction drum diameter is much larger than that of the yarn.

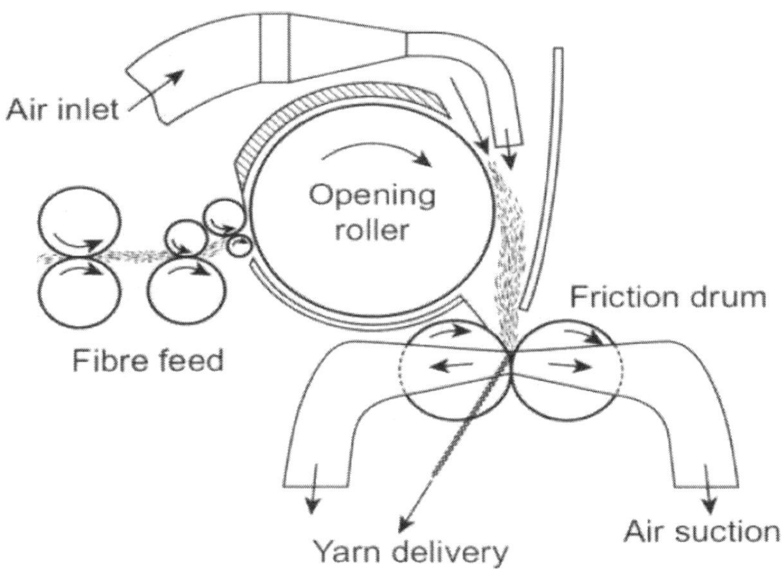

Figure 1.17 Schematic diagram on the Principle of friction spinning.

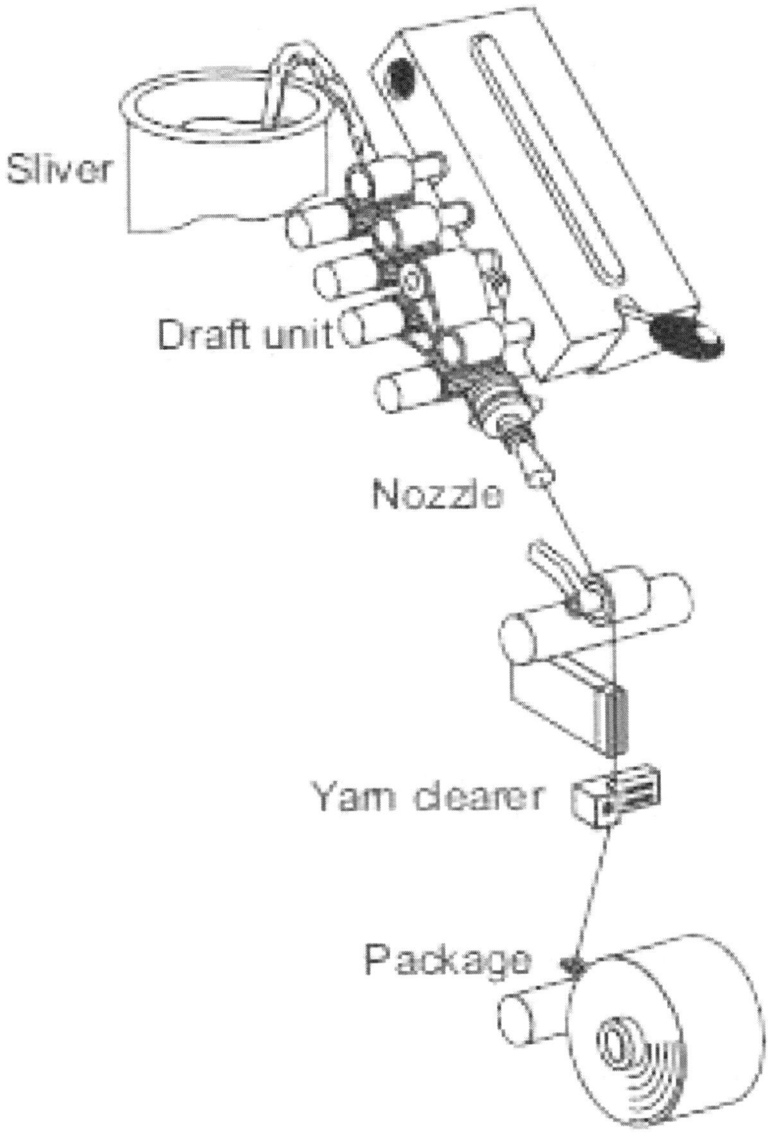

Figure 1.18 Schematic arrangement of the Air Vortex Spinning

1.8.5.3 Vortex spinning

Vortex open-end spinning technology has recently emerged as a revolutionary development in the field of fasciated yarn technologies. But the earlier impression of low quality and processing ability has hindered acceptance of these technologies. The most advanced fasciated yarnspinning technology that has gained momentum during the past decade is the Vortex-spinning technology (Figure 1.18). Vortex spinning was introduced by Murata machinery Ltd Japan at OTEMAS'97. This technology is best explained as a development of air-jet spinning, especially designed to overcome the limitation of fibre types in Murata air-jet spinning. The main feature of Murata Vortex spinning (MVS) is its ability to spin carded cotton yarns at speeds significantly higher than any other system currently in existence. The machine produces yarn at 400 m/min, which is almost 20 times that of ring-spinning frame production.

1.8.6 Air-Jet Spinning

This is a pneumatic method and is not an open-end spinning process. The drafted fibre strands pass through one or two air nozzles located between the front drafting roller and the take-up system (Figure 1.19). The roller drafting system drafts the input sliver into a ribbon-like parallel fibre strand. High-pressure air is injected into the nozzles, causing swirling airstreams inside the nozzle. This results in the insertion of false twist into the drafted fibre strands. The edge fibres wrap onto the surface of core strand and form the yarn.

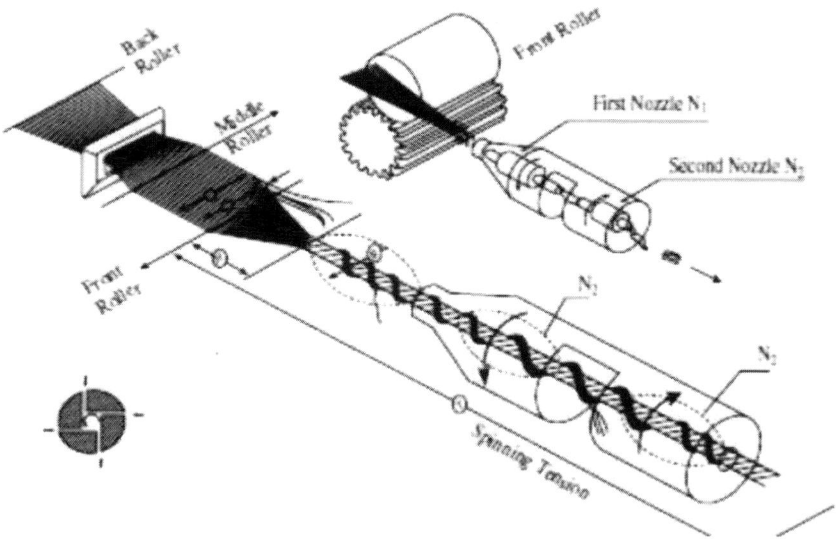

Figure 1.19 Air jet spinning

1.8.7 The Chenille Yarn System

A method of producing a chenille yarn that forms two ends at each unit is illustrated in Figure 1.20. The effect yarns are wrapped around a gauge or former which is triangularly shaped at the top, narrowing towards the base to allow the effect yarn coils to slide downwards onto the cutting knife. The width at the bottom of the gauge determines the effect length, by maintaining the depth of the pile, or 'beard', in the final yarn. Although, for the sake of simplicity, the cutting knife is shown in Figure 1.20 as a straight knife edge, modern machines all use a circular cutting knife.

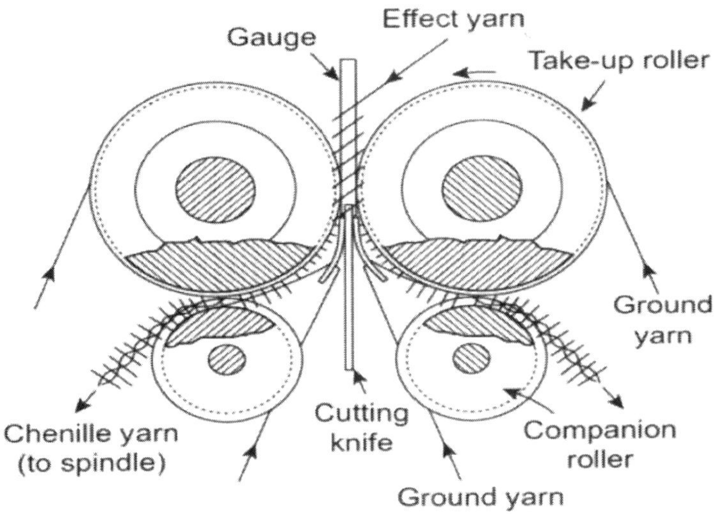

Figure 1.20 Chenille yarn production

On each side of the cutting knife there are two ground yarns, which may be either single or twofold yarns. One ground yarn is guided by the take-up roller while the other is guided by the companion roller. The take-up roller is pressed against the profiled guide and inter-meshes with the companion roller, allowing the two ground yarns to trap the pile created by the effect yarn in between them at right angles to the ground yarn axis. The two ground yarns are twisted together, usually by a ring spindle at the lower part of the machine, to produce the final yarn.

1.8.8 Flocking

Chenille effects can also be manufactured by a flocking process in which a ground yarn coated with adhesive is flocked electrostatically with loose fibres (Figure 1.21). The loose fibres and the ground yarn are charged with opposite electrostatic charges. As a result of this, the loose fibres are attracted to the ground yarn and are bonded to it by the adhesive. The loose fibres have the same electrostatic charge and repel each other, leading to good fibre separation and also forcing them to 'stand' on the ground yarn rather than lie flat on its surface. This is a very economical manufacturing method, but the yarn has poor abrasion resistance because the anchor of the loose fibres onto the ground yarn is weak and they can therefore by easily be worn off, leaving the ground yarn bare.

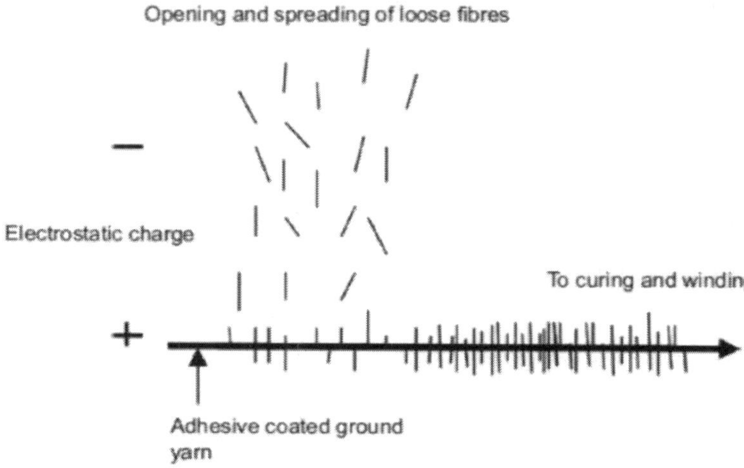

Figure 1.21 Chenille effect by flocking

1.8.9 Mock Chenille

A mock chenille effect can be manufactured by plying two gimp, boucle, or loop yarns with dense effects (two loop yarns with large numbers of small loops). The yarn may not look like chenille, but when it is made into fabric, the large number of small loops in the fabric results in a fabric surface resembling a chenille effect.

Chapter 2

Fibre to Yarn Staple Fibre Spinning Preparation

2.1 Introduction

Yarn production methods were originally developed for spinning of natural fibres including cotton, linen, wool and silk. Since the overall physical characteristics of the fibres and the mechanical processing factors involved differed from fibre to fibre, separate processing systems were developed. As synthetic fibres were introduced, synthetic spinning systems for texturized and untexturized cut staple were developed as modifications of existing staple systems, whereas spinning systems for texturized and untexturized filament were developed separately. Fibres can be classified as either short staple fibres or long staple fibres. This chapter will review the production of yarn from staple fibres. A staple fibre is defined as a noncontinuous fibre of relatively short length. Because of their short length, staple fibres must be twisted together to form a long, continuous yarn (hence the term 'spun yarns'). Staple fibres are usually natural fibres, although synthetic fibres can be cut into similar short lengths to be blended with natural staple fibres or used on their own to produce yarns with a natural feel.

Natural staple fibres include cotton, wool and flax. The natural staple fibres used in the textile industry are classified by the typical length of the fibre and are described as 'short staple' or 'long staple':• Short staple fibres have a maximum length of 60 mm. Cotton fibres are short staple fibres, having a length of 25–45 mm. Cotton linters (the fibres that remain adhered to the cotton seed after the first ginning) are even shorter, at just a few millimetres long; these are used for the manufacture of lower-quality products such as lint and cotton wool.• Long staple fibres have a length of more than 60 mm. Wool fibres are long staples, with a length of about 60–150 mm. As a result of their short length and non-uniform nature, staple fibres require greater processing before a satisfactory yarn can be produced; this obviously adds to production costs. However, the desirable characteristics of staple fibre yarns – such as comfort, warmth, softness and appearance – often compensate for these increased costs.

The system used to prepare staple fibres for spinning depends on the type. Short staple fibres are prepared using the cotton system, while longer wool fibres may be prepared using either the woollen system or the worsted system (also called the long staple system). It should be noted here that shorter wool fibres (<40 mm) can also be processed using the cotton system. These systems will be discussed in turn in the following sections. The chapter will then review the various different spinning techniques that are used to produce staple fibre yarns and the characteristics of the yarns produced by the different methods. Table 2.1 gives the details of operations and machines involved in spinning.

Table 2.1 Details of operations and machines of spinning

Operation	Machines
Opening	• Blowroom machines • Card • OE rotor spinning machine
Cleaning	• Cleaning machines • Card • Comber • Drawframe (dust removal) • Rotor spinning machine
Blending	• Blowroom machines • Card (fibre blending) • Drawframe
Aligning	• Card • Comber • Drawframe • Roving frame • Final Spinning machines
Uniting	• Card • Comber • OE Rotor spinning machine
Equalizing	• Card with leveller • Drawframe • OE Rotor spinning machine
Attenuating	• Card • Drawframe • Roving frame • Final Spinning machines
Imparting Strength	• Final spinning machines
Winding	• Roving frame • Final spinning machine

In recent years a number of modern staple spinning processes other than ring spinning have been developed that reduce or shorten the number of steps necessary for formation of yarns suitable for textile substrate formation and are discussed separately following conventional ring spinning techniques. Yarn preparation from fibre filaments is much less complex and often no or only limited twist is imparted prior to use in the textile substrate

2.2 Preparation of Cotton and Other Short Staple Fibres

Cotton fibres arrive at the cotton spinning mill in the form of cotton bales that are packed densely, typically with wrapping secured with polypropylene ties. At this stage, the cotton contains 1–15% impurities (e.g. dust, dirt, vegetable matter) which must be removed as the cotton is processed. In order to convert the raw cotton into fibres that are separated and aligned in a suitable manner for yarn production, the cotton has to pass through the following processing stages:

1. Opening and cleaning 2. Blending 3. Carding 4. Combing (optional) 5. Drawing 6. Roving

Stages 1 to 3 take place in the blowroom, so called because the cotton is transferred from stage to stage using pneumatic transport, i.e. it is blown from one machine to the next. The fibres that leave the blowroom are separated into disentangled individual fibres which can then be further processed ready for yarn production.

We may consider the first three stages collectively as material preparation. This chapter is concerned with the process technology of stage I. In the conversion of baled cotton into finished yarn, the primary purpose of the preparatory processes is to open, clean, and parallelize the fibres and then present the material for spinning. In doing so, these processes convert a threedimensional bale of compressed, entangled, matted fibre mass into an orderly arrangement of fibres in a one-dimensional continuous strand length. The objective is for the conversion to be achieved with minimal fibre breakage and no fibre entanglement remaining in the strand length.

A great deal of attention has been paid to, among other factors, improving machine setting and the operating speeds of component parts so as to attain gentle working of fibres and to avoid fibre breakage.

Fibre properties (such as length, fineness, strength, elongation, frictional characteristics, and the level of impurities present in the fibre mass) are of much importance to machine design and development. Man-made fibres (mmfs) present little problem where impurities are concerned, and, compared with natural fibres, their properties can be more easily modified to meet process requirements. Therefore, the development of most production machines has largely focused on cotton and wool processing; this is particularly so in the area of material preparation.

2.3 General Principles of Short Staple Spinning

The short staple yarn spinning systems make use of the following basic principles:

(1) Mixing

The term mixing is applied for combining or blending more than one mass of fibres into a single homogenous mass. Since all the natural fibres vary from one another from region to region, field to field and even from bale to bale, it is necessary to mix them well so that the final end product i.e. the yarn may be homogenous and of consistent quality.

Fig. 2.1 Mixing

(2) Blending

Usually blending refers to the mixing and combining two or more than two fibres of different origins. For example cotton fibres with polyester. The purpose of blending is to achieve desired properties of both the components of the fibres in a yarn which otherwise is unobtainable from a single component. For example strength, crease resistance, colour, price, moisture absorption, drying properties, etc.

Just as mixing, blending should also be done thoroughly so that a homogenous yarn may be produced out of the component fibres.

(3) Opening

The raw fibres for easy transportation are highly compressed into bundles called as bales.

Before effective spinning can be carried out, this compressed form of fibres must be separated from one another. Opening refers to mechanical separation of fibres from one another. The opening is carried out gradually in steps. In the first step the bale of fibre is broken down into small tufts then later on at various stages of spinning these tufts are further broken down into individual fibres.

(4) Cleaning

All the natural short staple fibres have some amount of natural impurities in it. These impurities have to be removed before fibres are converted into a yarn. On the other hand the short staple manmade fibres just require opening but no cleaning. The amount and type of cleaning required for a particular lot of fibres depends upon size and amount of impurities present in it. Greater the trash content in fibres greater number of cleaning points are used during various stages of the spinning process. Typically cotton bale contains about 5 to 10% impurities and about 3.5 to 7 % cleaning is obtained during initial opening and cleaning. Cleaning is defined as the process of the removal of unwanted trash by mechanical actions. Natural fibres such as cotton will inevitably contain impurities such as leaf, seed, trash and dust, which must be removed if high-quality yarns are to be produced. The opening and carding operations of the blowroom machines disentangle these impurities from the fibres, which can then be removed using air currents that simultaneously transport the cotton tufts. During the process of breaking up the fibre mass into initially large tufts, these large tufts into smaller tufts, and so on, fibres of one part of a tuft bundle slide past fibres of the other part. Light particles of impurities, such as dust, are freed and can be removed by air currents. Larger particles of leaf, seed, dirt, and sand that are lodged between fibres are loosened, and some are sufficiently freed to be removed by beating the tufts against grid bars or perforated plates. We may refer to these actions as mechanical cleaning actions or simply cleaning. It can be understood that the opening action leads to both cleaning and reduction of tuft size. Invariably both opening and cleaning take place simultaneously.

Cleaning is generally associated with opening. As the fibre opening takes place, the impurities also get separated and removed. Greater the degree of opening, greater cleaning may be carried out. The fibre separation is the further opening of the small tufts that have been initially opened up into an individual fibre form. This happens at the carding machine. By fibre separation a further up to 1% waste can be removed.

(5) Fibre Regularising

Fibre regularising means the reduction in the variation of the mass per unit length of the fibre material. Regularising is very important to produce a regular and a more consistent quality of a yarn. Fibre regularity takes place at the drawing frame.

(6) Fibre Alignment

Along with the improvement of the fibre regularity, the alignment of the fibres also improves at the draw frame. Fibre alignment means the straightening of the fibres and making them parallel to the axis of the sliver.

(7) Drafting

Drafting is drawing out or attenuation of the fibre material into a desired fineness or count of a yarn. In ring spinning system drafting is done in two stages one at the roving frame and other at the ring frame.

(8) Twisting

Twist plays a vital role in the final formation of the yarn. The purpose of giving twist is to bind the fibres together as a yarn and to give strength to the yarn. The major twisting takes place at the ring spinning frame or at the final stage of spinning.

The impurities in cotton fibres can be classified into:

Seed: Seed impurity is the largest type of impurity present in raw cotton and it includes unginned seeds with fibres attached to it, ginned seeds and parts of seeds.

Chaff: The chaff is the vegetable fragments consisting of leaf particles, bract, shale and stalk of the cotton plant. Bract is a small type of a leaf that grows beneath the cotton boll and shale is the silvery interior lining of the cotton boll.

Dirt: The dirt impurity includes soil and sand particles that may be added from the cotton fields due to mishandling of the fibres and also the cotton fibres pick up dust and sand if they are transported by open trucks.

Micro-dust: The micro-dust includes very fine particles of chaff, dirt, small fibre fragments and spores of mildew. These particles are extremely small and are often a fraction of the fibre diameter. They generally gets embedded around the natural wax of the cotton fibre.

Abnormal Impurities: The abnormal impurities are very rare however when found in the cotton fibres, they can cause serious problems. These impurities include pieces of stones, pieces of iron, cloth fragments, foreign fibres such as jute, polypropylene, etc. (that may be included due to the bagging of cotton fibres made up of these fibres), grease and oil (from machine harvesting or ginning), tar and coal (from the air while the cotton fibres are transported openly), small pieces of wood, etc.

The term trash is applied to all of these impurities present in the raw cotton. The total trash content of cotton fibres ranges from 1% to 10% of the total weight of the cotton fibres. The amount of trash content directly determines the amount of cleaning required. The removal of impurities is associated with the opening of the fibres and is carried out in stages during the spinning process. Initially the impurities are removed in the blow room with the very basic opening of the larger tufts of fibres into smaller tufts. Then further at the carding machine due to fibre separation, more impurities are removed. The blow room mostly removes the seed and chaff while the carding removes the dirt and microdust. Greater the impurities present in fibres more opening and cleaning is required. However with more severe opening and cleaning, expensive fibres can be damaged or they can be lost during the process.

2.3.1 Opening and Cleaning

The early stages of material preparation generally involve the removal of impurities from thefibre mass by mechanical or chemical means and the blending of the fibrous mass to produce a homogeneous feed to the next stage. Several machines are usually arranged in sequence to carry out the cleaning and blending of the fibre mass. The type of machines and the sequence used will depend on the fibre type and grade to be processed. We first concentrate on the fundamentals of cleaning and blending, giving examples of machines employed for such tasks, and then present an overview of a typical machine sequence for short-staple, worsted, and woollen systems. Figure 2.1A shows the various operations that occur in the blowroom. The term 'opening' refers to the process whereby the high-density cotton sheets from the bale is broken down into large clumps or tufts of fibres; these tufts are subsequently further broken down into smaller tufts ('tuft lets'). The action of progressively breaking up the fibre mass into smaller clumps is referred to as opening. Opening is defined as the process of breaking up of the fibre mass into tufts. At this stage the individual fibres are not separated. In modern production mills, numerous cotton bales are unwrapped and arranged in lines in the blowroom for the start of processing. The automatic bale opener, which consists of sets of opposing points or spikes, that works its way along a line of bales (termed 'bale lay downs'), plucking tufts from each bale.

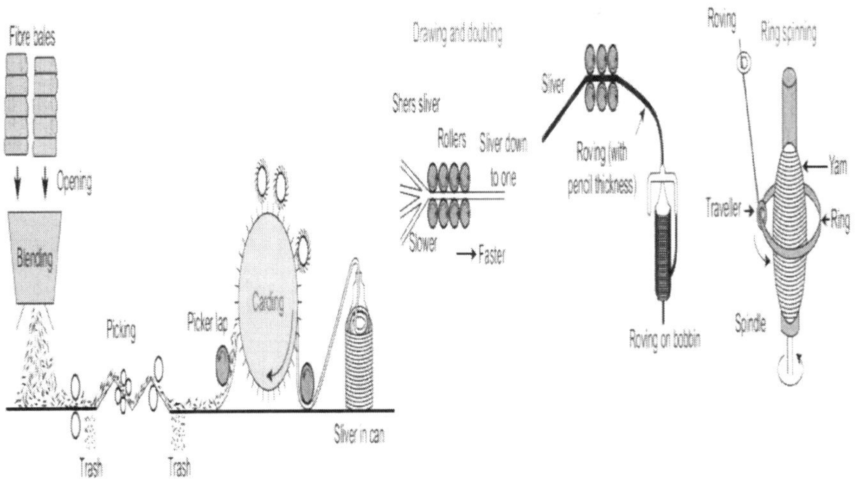

Figure 2.1A Various processes of short staple spinning

The tufts are then transported pneumatically to the next stage, which is the cleaning operation. At this stage a 'pre-mixing' operation is also carried out in which bales of different cotton grades or even fibre types (cotton + viscose, etc.) can be positioned in the bale lay-down to achieve a certain blended specification.

Because of random variations, fibres from differing regions of the same bale, as well as between bales of the same batch of raw material, will differ in properties, and the difference is more marked for natural fibres than for manmade fibres. Therefore, it becomes necessary to mix, the fibre tufts obtained from opening the various bales to be processed as thoroughly as possible to have a uniform end product.

Tuft blending is defined as the process of mixing of fibrous tufts obtained from opened bales to produce a homogenous mass of fibrous material with an aim to produce consistent yarn properties. Tuft blending may take place from the start of removing tufts from the baled fibre mass, but machines exclusively designed for blending are incorporated within the sequence of opening and cleaning machines so as to use a suitable tuft size for ensuring optimal blending.

Often, different grades of a natural fibre or different fibre types are blended for improved product economics, product performance, or both. Tuft blending is therefore an integral and essential part of the early processing stage.

Each machine that opens and cleans the fibre mass is referred to as a cleaning point, although some machines do not reduce tuft size but only expel dust, dirt, and trash particles from tufts. Machines are usually placed sequentially in a cleaning line of a blowroom so as to progressively intensify the degree of opening and cleaning and to blend the tufts. The fibre tufts are usually transported from one machine to another by airflow through connected fibre transport ducting. At the end of the cleaning line, 40 to 50% of impurities are removed (largely heavy particles), and the opened material is then fed into the carding process.

Opening and cleaning machines operate and function on one or more of the following principles/actions:

- The action of opposing spikes, which is basically an opening action
- The action of beater and grid bar, is responsible for both opening and cleaning
- The action of air currents, leads only to cleaning

Methods of Opening & Cleaning: The mechanical techniques or methods used for the opening and cleaning of the cotton fibres are:

Spiked Surfaces

This action consists of plucking and gripping the mass of cotton fibres between two sets of spiked surfaces moving in opposite directions. This action opens the fibres and does some cleaning as well. The schematic diagram showing three types of spikes is given in Figure 2.2.

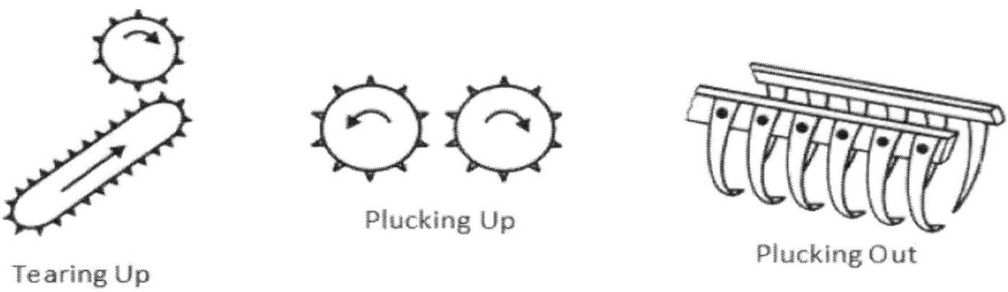

Figure 2.2 Spiked Surfaces

Beaters

The term beater in spinning is generally referred to a revolving drum having blades, spikes or pins mounted on its surface. The fibres are fed and subjected to the surface of the beater with the help of air currents or by using pair of rollers. When the surface of the beater strikes the fibre mass, it plucks or tears the fibres and transports the opened fibre to next stage. During this opening, the trash being heavier falls down and is collected below the machine. Figures 2.3 and 2.4 give the schematic representation of three types of beaters.

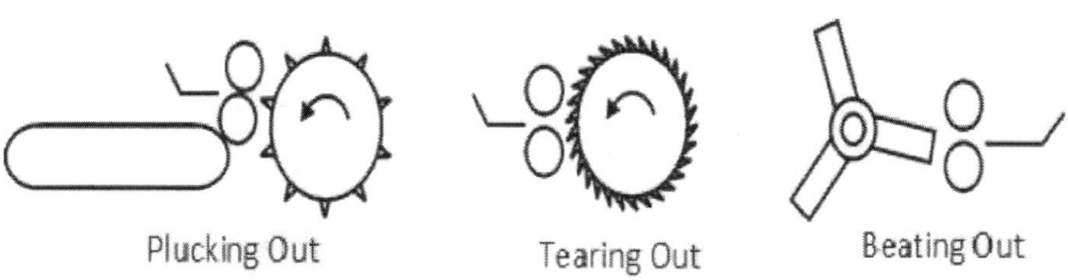

Figure 2.3 Beaters

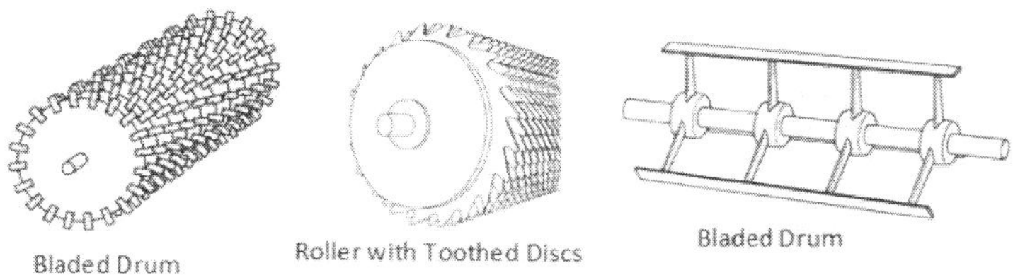

Figure 2.4 Beaters

Air Currents (Pneumatic Action)

Lighter impurities like dust and micro-dust cannot be effectively removed from the fibres by using either the spiked surfaces or beaters. These impurities can be removed effectively by passing the fibres over the perforated grids where the air currents can draw off the dust mingled with the fibres. Fibres being buoyant are easily carried up by the air current whereas the relatively heavier impurities fall off due to momentum or gravity. The diagrammatic representation of the same is given in Figure 2.5 All the above and majority of the operating devices in the blow room machinery are functioning as opening devices when they are used alone. However, these very same opening devices can also work as cleaning devices if they are worked in cooperation with cleaning apparatus such as grids and special machine covers.

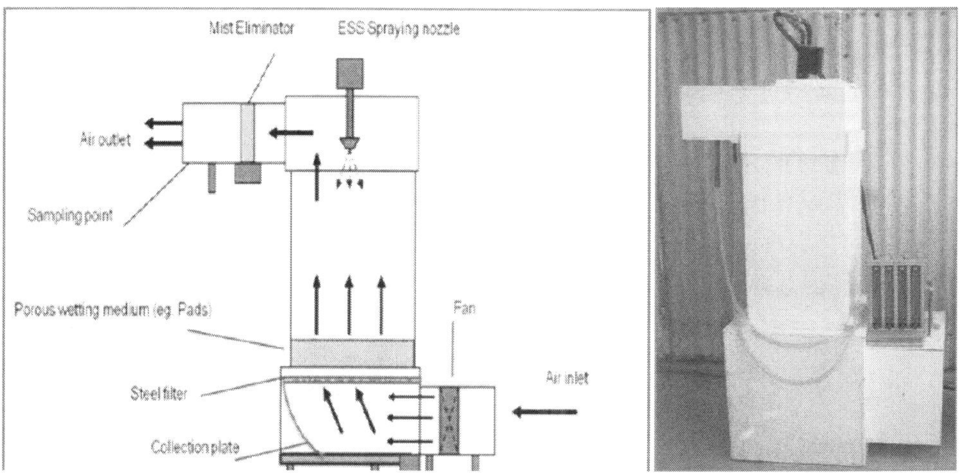

Figure 2.5 Schematic diagram of Fine cleaning using air currents

The following elements can be used in the grids of the blowroom:

Slotted Sheets

Perforated Sheets

Triangular Section Bars

Angle Bars

Blades

These elements can either be used alone or in combination with each other. For example, the perforated and slotted sheets were widely used and are placed under the cards. But, now their use is diminishing. Same is the case with the blade grids that were used in combination with triangular bars grids. Figure 2.6 shows the diagrams of these elements.

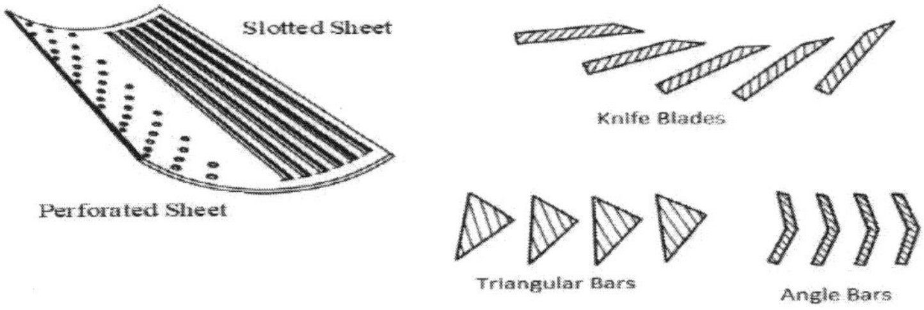

Figure 2.6 Schematic diagrams of the different elements of the grid.

Even though, the angle bars are very robust, they tend to create fibre blockages. Therefore, these days, in the modern cleaning techniques, either the triangular grid bars or knife blades are used alone for the most efficient waste removal and gentle treatment of the fibres. The impurities and the fibres that fall through the gaps of the grids and are collected in a special waste chamber of the machine which is placed under the grid. In old machinery this waste has to be manually removed periodically. However, in the latest opening and cleaning blowroom systems pneumatic waste removal systems are invariably used.

Grid Adjustments

The grids can be in one, two or three parts so it can be adjusted as a unit or in individual sections. The following three basic adjustments are possible.

1. Distance of the complete grid to the beater.
2. The distance or the gap between the adjacent grid bars
3. Angle relative to beater blade

It is seldom possible to make all these three adjustments simultaneously, generally machines are so designed that only two adjustments are possible at a given point of time.

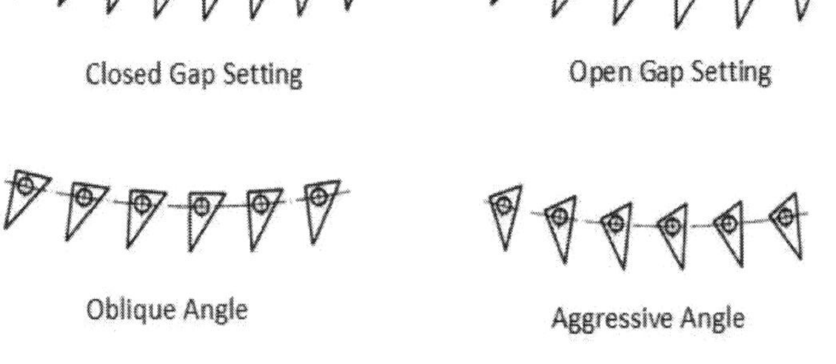

Figure 2.7 Schematic diagrams of the different settings of the grid bars

Cleaning Efficiency

All openers & cleaners along with foreign matter also tend to remove a certain amount of short fibres (also called as lint). The collected mass of trash, dust, fibre fragments and fibres removed is called as waste. The amount of waste removed at each stage of cotton cleaning stage should be known for evaluation.

2.3.2 Factors Influencing Opening & Cleaning

The degree of opening, the degree of cleaning and the fibre loss are primarily dependent on the following factors:

The type of opening devices used.

The speed of the opening devices.

Type of feed

Degree of penetration.

Spacing of feed from the opening devices.

Type of grid used.

Area of grid surface.

Grid Settings.

Airflow through the grid.

Condition of pre-opening.

Position of the machine in the machine sequence.

2.3.3 The Blow Room

The blow room is the first stage or the first process in the short staple mechanical spinning processing sequence. The name 'Blow Room' is given to this stage because of the air currents that are commonly used for blowing the fibrous mass and dust during the processing of fibres. In the blow room the fibre mass is progressively opened, cleaned and mixed. This is done by using a large number of machines. In each of these machines used in blow room if the actions are too severe of sudden then fibre damage will occur.

Functions of Blow Room

Fibre Opening: The tightly packed fibre bales received from the ginning mills have to be opened by opening out the larger fibre lumps into smaller tufts and ultimately opening out the smaller tufts into individual fibres.

Cleaning: All the natural fibres including cotton have considerable amount of impurities present along with them which have to be removed in order to produce a clean yarn.

Mixing/Blending: The properties of the cotton fibres differ from each other from bale to bale and from fibre to fibre because of the natural variations. Therefore, in order to obtain a homogenous and consistent quality yarn, these fibres need to be thoroughly mixed together. Sometimes, in order to have a desired quality at the right price high quality cotton fibres are purposely mixed with low quality fibres.

Preparation of feed for the next stage: The end product of the blow room that is the blow room output should be compatible with the next stage of spinning i.e. carding. The feed to the carding can be either given in the form of lap or in direct fibrous tufts through the chute feeding system.

2.3.4 Blow Room Machinery

In the blow room, all the above functions have to be achieved efficiently and effectively without much wastage and fibre damage. For this purpose various different types of machines are used.

Based on the nature of functionality of these machines, they are classified into five types:

1. Bale Opening Machines
2. Mixing Machines
3. Cleaning Machines
4. Dust Removing Machines
5. Recycling Machines

For these machines to perform optimally, they must be located at specific positions in the blow room line. The modern cotton blow room line can be divided into six distinguished zones based on this purpose.

1. Bale Opening Zone
2. Coarse Opening & Cleaning Zone
3. Mixing/Blending Zone
4. Fine Opening & Cleaning Zone
5. Intense Opening & Cleaning Zone
6. Card Feed Preparation Zone

In case the cotton to be processed has less impurities, then the Zone 5 (intense opening & cleaning) may not be necessary. Furthermore, the Zone 4 (fine cleaning) or the Zone 5 (intense opening & cleaning) and the Zone 6 (card feeding) can be combined together to form a single unit.

The additional operation of dust removal is not associated with any single zone; in fact dust removal is carried out at a greater or lesser extent at every machine of the blow room.

Bale Conditioning

The cotton fibres from the ginning factories are received by the spinning department in form of highly compressed bales that are wrapped by metal or plastic straps. As the bales are received these straps are cut off and the bales are allowed to be conditioned for 24 hours in controlled temperature and between 60% to 80% relative humidity. For this purpose all the spinning mills specifically allocate some floor space for conditioning purpose.

This conditioning allows the fibres to relax and maintain a temperature and moisture equilibrium with atmosphere and this reduces the chance of fibre damage and inconsistency in quality.

Bale Lay Down

After the bales are conditioned properly, they are taken to the blow room and placed in groups of special sequence for the first processing stage. This group of bales thus positioned is called as bale lay down. The main aim of a proper bale lay-down is to have fibres with similar properties and spinning attributes from lay-down to lay- down so that spinning machines can be adjusted optimally so that a yarn of consistent quality can be produced over a longer period of time.

2.3.4.1 Zone-1 (Bale Opening Machines)

After proper bale lay-down, the initial bale feeding is done either by using a parallel set of machines called as hopper feeders or by a single machine called as a bale plucker. The hopper feeders have stock supply compartments which are either filled manually or by automatic machines. A cross-sectional diagram of a typical hopper feeder is shown below:

(1)Feed Table (2) Internal Feed Lattice (3) Light Barriers (4) Baffle Plate (5) Brush Rolls (6) Spiked Lattice (7) Cleaner Roller (8) Evener Roller (9) Stripper Roller

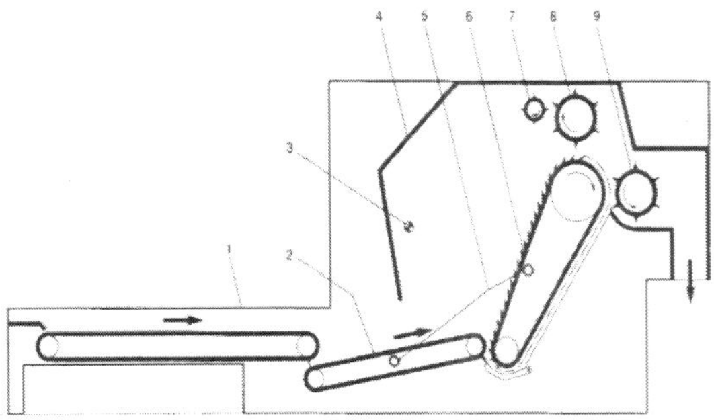

Figure 2.8 Cross-sectional diagram of a hopper feeder.

2.3.4.2 Yarn Manufacturing Process

Flow Chart of Yarn Manufacturing Process

MIXING
↓
BLOWROOM
↓
CARDING
↓
DRAWFRAME (BREAKER)
↓
DRAWFRAME (FINISHER)
↓
SIMPLEX
↓
RING
↓
AUTOCONER

Fig. 2.9 **Flow Chart of Yarn Manufacturing Process**

Flow Chart of Machine in Blow Room

AUTOMATIC PLUCKER

⇓

AUTOBLENDER WITH CONDENSER

⇓

STEP CLEANER

⇓

UNICLEAN B11

⇓

MULTIMIXER

⇓

DUSTEX

⇓

PINBEATER WITH CONDENSER

⇓

RESERVE HOPPER FEEDER WITH CONDENSOR

⇓

SCUTCHER

Fig. 2.10 Flow Chart of Machine in Blow Room

2.3.4.3 Automatic Plucker

This machine is used in the first stage of blow room and the purpose of this machine is to pluck the cotton from the stack of cotton layers with the help of beater. Then, by means of condenser these fibers are delivered to the next machine for further processing via a delivery pipe.

Fig.2.11 Automatic bale opener

2.3.4.4 Automatic Plucker

This machine is used in the first stage of blow room and the purpose of this machine is to pluck the cotton from the stack of cotton layers with the help of beater. Then, by means of condenser these fibers are delivered to the next machine for further processing via a delivery pipe.

2.3.4.5 Condenser

This machine is used for opening and transferring the various cotton fibers with the aid of the suction and eliminates the trash and linters in the material.

2.3.4.6 Auto Blender

From his name it is clear that this machine is used for blending of different cotton fibers. After stack mixing the material is fed to this machine. It consists of two types of lattices i.e. feed lattice and spiked lattice and one U-type spike beater as well. The material is fed from the feed lattice and then the spiked lattice takes this material to the beater for beating process which opens the material. Due to material opening cleaning also takes place. After beater again there is a spiked lattice, which takes the material and fed it to next machine.

2.3.4.7 Step Cleaner

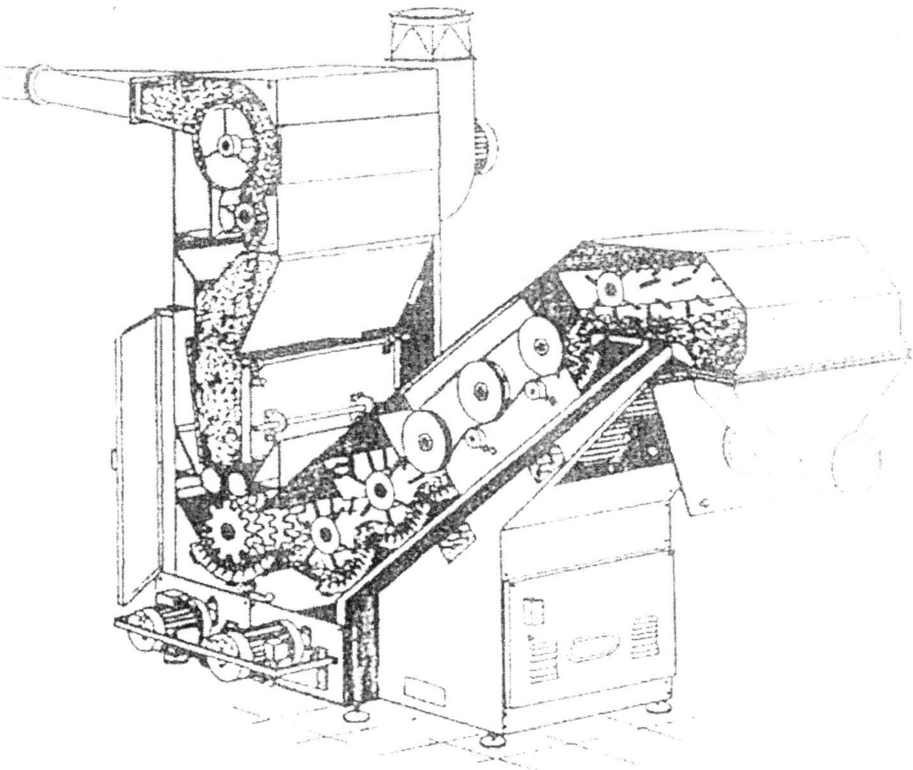

Fig 2.12 Step cleaner

This machine consists of 6 inclined beaters as shown in the figure. The main purpose of this machine is opening and cleaning. First material is pneumatically fed to the first beater, which is at the lowest level. It beats the material against grid bars and finally sends it to the second beater and so on. The beaters speed gradually increases from first to the 6th beater so gradual opening of material takes place in this machine. When the material strikes on the grid bars it opens and waste is extracted through the grid bars whereas the material is transferred to the next beater.

2.3.4.8 Uniclean B11

Figure 2.13 Uniclean B11

Working principle

This machine is designed for intensive cleaning. The material is fed to the single opener and passed mechanically six times over the cleaning grid on special points. In this process the raw material is guided over the integrated dedusting filter where dust, fiber fragments and trash are stripped off mechanically. The subsequent process stages have to remove only a small proportion of the trash while it can removed up to 70% tarsh. (Engineering Manual of Uniclean B11, Rieter, Switzerland)

2.3.4.9 Multi Mixer

In this machine the process is done in different chambers to get the maximum mixing of the raw material. It is also used to form different blends of cotton and polyester or any other fibers.

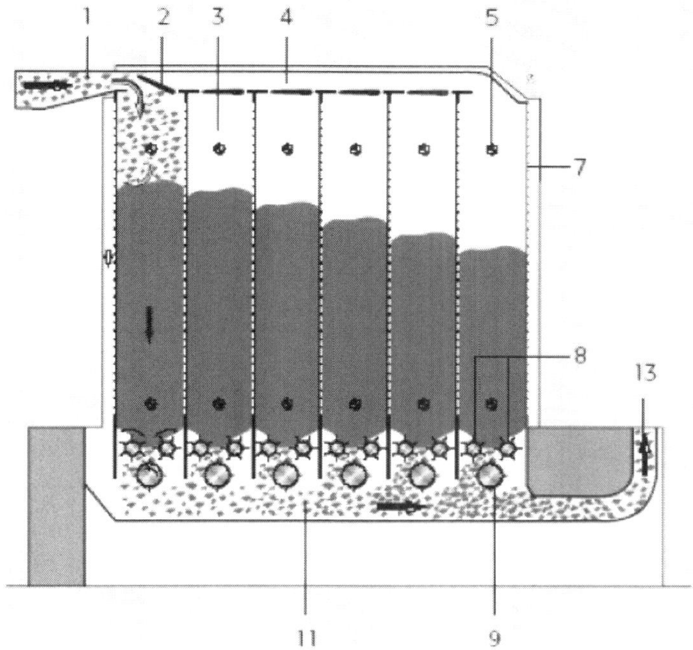

Figure 2.14 Multi mixer

Working Principle

- ❖ The last chamber is filled firstly and then the rest of the chambers.
- ❖ When the chambers are full the delivery rolls are activated underneath each chamber and the material is transported to the conveyor by the opening roll or beater. The beater is moving at a nominal speed.
- ❖ When the chambers are empty they are again filled by the condenser.
- ❖ A small amount of tufts from the several chambers builds up on the blending conveyor.

After that the material is fed to the cleaner or opener.

2.3.4.10 Micro Impurities Deduster (DX)

Dedusting is another important process apart from the opening and cleaning of the raw material. A good dedusting result in higher efficiency rates, less yarn breakage in winding and knitting process.

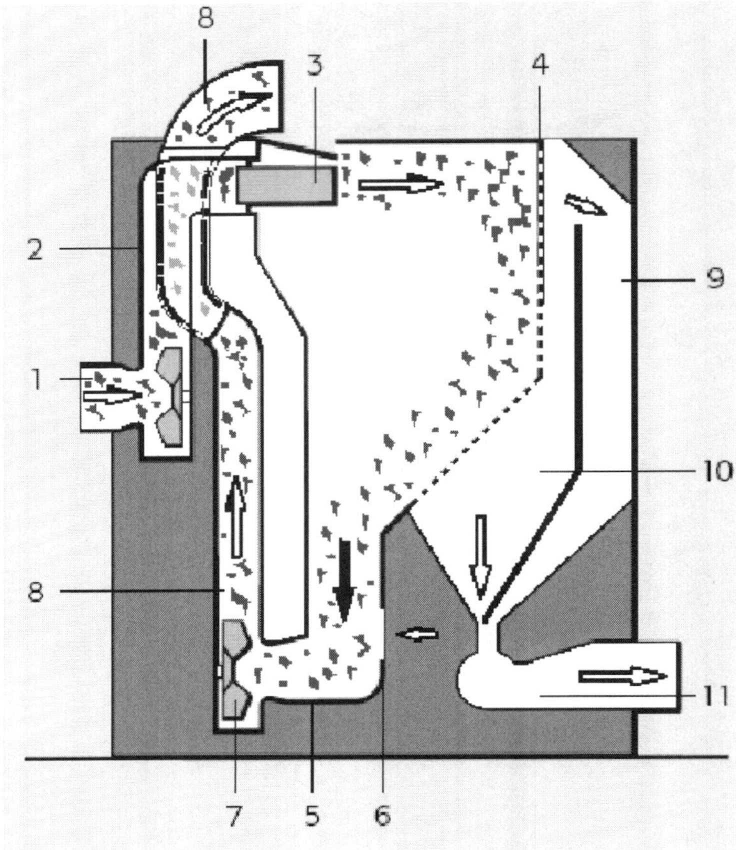

Figure 2.15 Micro impurities deduster (DX)

This machine is suitable to deal with various grades of cotton and can remove part of any tiny waste, micro-dust as well as short fibers away from the opened fiber.

2.3.4.11 Pin Beater

It is a fine cleaner and gives sever opening action to the material so a nip feeding is provided here. Material is fed to main beater, which consists of lot of pins which beats the material against the grid bars at very high speed so provides an intense opening action. As the material opens, the waste is extracted through grid bars. The intensity of cleaning depends upon the grid bars spacing and the gauge between pins and grid bars. Material is fed to the machine pneumatically through condenser and hopper.

2.3.4.12 Double Hopper Feeder

This m/c is inserted in front of the electrical distributor and behind the single beater Scutcher and lap m/c also in connection with the condenser. The raw cotton, destined for carding and combing system are to be processed in this m/c into cotton fleece and then fed into single beater Scutcher and lap m/c to produce laps.

2.3.4.13 Scutcher

Figure 2.16 Scutcher

It is the last machine of Blow room department. The function of this machine is to make end product in the form of lap for carding department. Lap must be wound by the winding rollers on a lap tube. In order to ensure a large take up capacity, the lap tube is pressed against the winding rollers. A lap weighing device connected with the lap apparatus detects deviation of lap weight from asset valve.

2.4 Carding

2.4.1 Introduction

The blow room process opens the compressed bales of cotton into small tufts and also removes 75% trash present in the cotton. Still the cotton has to be opened to individual fibre state and it must be free of impurities and trash etc. up to the maximum possible extent so, the fibre-to-fibre opening and complete removal of trash is carried out on the card machine. It is therefore rightly said that "*Half Carded, Half Spun*" and "*Well Carded Well Spun*" because it is the process of carding which plays a very important role in the spinning process and is called as the *Heart of Spinning*..

Following are the main tasks of the card machine:

- Opening of fibers to individual state.
- Elimination of impurities present in cotton that were not extracted in the previous cleaning departments.
- Disentangling of the neps
- Removal of the short fibers
- Fibre blending
- Fibre orientation
- Sliver formation

The production of card deptt is 263 bags/day. .

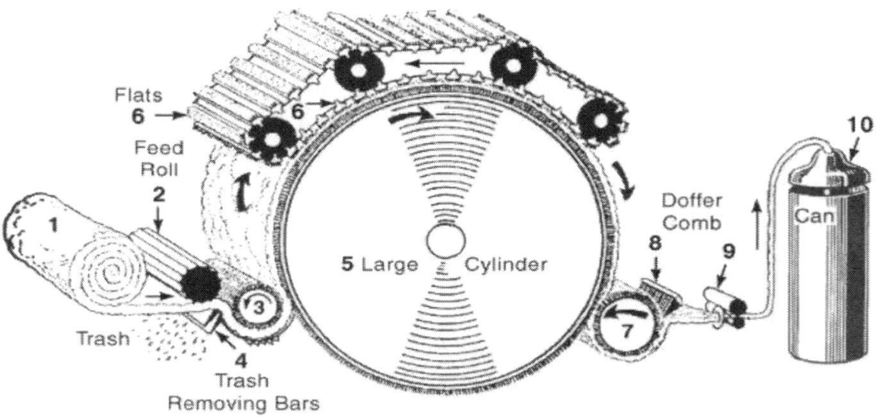

Figure 2.17 Sectional view of card

The process starts with the output of blow room department which is called Lap (1). The lap from blow room is unrolled and placed on the back of feed rollers (2), and then feed rollers transferred this material to taker in (3) which has wires like saw teeths. The function of taker in is to shift the lap against trash bars (4), which extracts trash particles from the cotton, and then passes these fibers into the cylinder (5). The body of cylinder consists of thousands of fine wires which grasps the fibers. The flats (6) are mounted on the upper side of cylinder having more fine wires, and they are moving in the opposite direction as compare to the cylinder. The short fibers are removed by the flats and long fibers remains on the cylinder which then transfers to doffer (7). On the top of doffer, doffer comb (8) is placed which vibrates against the doffer and take away the floating fibers from it. Now, fibers in the form of web, transfers through condenser rolls (9), where fibers are condensed and then moves into a can through a coiler head (10) in the form of sliver .

2.5 Draw Frame

Figure 2.18 Draw frame

2.6 Introduction

Draw Frame is a machine on which drawing is done. In this section sliver from card deptt, is again processed. For this purpose two types of passages are done to that sliver with the help of breaker and finisher drawings. The reason behind these two passages is to make sliver more uniform, even and parallel.It contributes less than 5 percent to the production cost of the yarn but its influence on the yarn quality is enormous.

The production of drawing section is 254 bags/day. .

Tasks of the drawing frame

- ❖ Improving evenness
- ❖ Blending
- ❖ Parallelization
- ❖ Dust removal

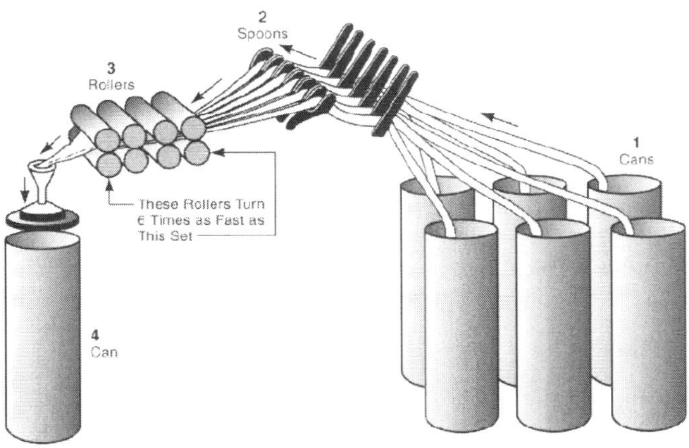

Fig.2.19 Sectional view of drawing Frame

Working principle

From the carding machine, cans (1) are fed to the drawing frame then, the sliver were taken out from the card cans. Each sliver is taken out and guided to the nip of back roller (3) with the help of spoons (2). Drafting is done with the help of difference of speed of rollers. After drafting, the material is in the form of ribbon. With the help of trumpet guide ribbon is accumulated and condensed to form a circular shaped sliver. Furthermore, this sliver is collected in the cans (4) for further processes .

2.7 Roving Frame or Simplex

2.7.1 Introduction

"Roving frame is known as necessary evil". The problem is that this machine has many complications, creating faults which have no solution, increasing production costs, its finished product is sensitive while using in winding and unwinding operations. Then, why this machine is still in use?

It has two reasons:

1. As far as the structure of drawing sliver is concerned it is thick and hairy and creates fly during working. Draft needed to convert this thick sliver into a yarn is 300 to 500 so, it is not possible for ring machines to make a yarn on this single drafting system that fulfill all the demands.
2. Presentation of feed material to the ring spinning department and draft can signify the worst conceivable mode of transportation.

Fig.2.20 Roving frame

Tasks of Roving Frame

- ❖ Attenuation-drafting the sliver into roving.
- ❖ Twisting the drafted strand.
- ❖ Winding the twisted roving on a bobbin.

The production of this department is 219 bags / day. .

2.8. Ring Frame

2.8.1 Introduction

From the early ages of history, the production of yarn has been done by spindles. During the last centuries many spinning machines came out but no one has been able to replace the ring spinning machine.

Advantages of ring spinning frame

- ❖ We can make every count of yarn in this machine

The material obtained from this machine has optimum characteristics, especially if we talk about structure and strength

- ❖ Very simple and easy to monitor
- ❖ Very modernized, less efforts required and accessible for everyone.

In case of fine counts it is usually 200 bags/day while if course counts will be processed it reaches upto 400 bags. .

Tasks of Ring Spinning Frame

Drafting

The purpose of drafting is to get the desired thickness.

Twisting

The twist has its own importance for inserting the strength in the yarn because twist interlocks the fibers with one another..

Package formation

At the end the yarn is wound on small packages (ring bobbins) and stores it for further processes.

Working principle

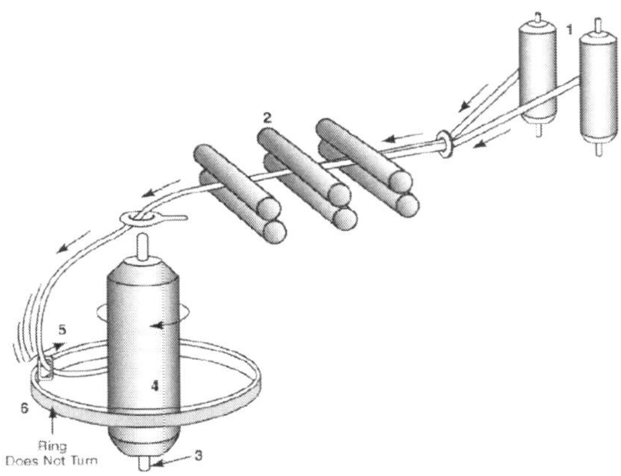

Fig.2.21 Sectional view of Ring Spinning Frame

The process starts from the output of simplex machine which is called "roving". Roving is wound on roving bobbins (1). With the help of guides this roving is fed to the pair of drawing rollers (2) which draws this thick strand to its final desired count. The ring bobbin which is inserted into the spindle (3) is moving with the constant speed. The speed of first two pairs of drafting rollers is almost same but the final pair rollers speed is adjusted according to amount of yarn delivers so that it is twisted by the desired quantity as it is wound on the bobbin. The yarn is directed by the traveler (5); the traveler imparts twist to the yarn and is present on the ring (6) surface around the bobbin. Due to yarn insertion from traveller it drags and yarn winds on the bobbin with the same speed as the front rollers. .

Fig.2.22 Traveller

The traveller's rotational speed is less than the spindle speed by reason of the frictional force produced when it slides on the ring surface and one more factor is the air resistance due to the motion of yarn between the yarn guide and the traveller.

2.9. Winding

2.9.1 Introduction

The basic function of winding machine is the transfer of yarn from smaller packages (ring bobbins) to bigger package (cone), which is suitable for transport and for further processing. Suppose if we use bobbins from ring directly into weaving or knitting process, then we got following problems:

The bobbins will be finished very quickly so more time will be wasted in bobbins replacement while changing the new with the old ones.

- While changing bobbins again and again it results in the form of many defects in the fabric.
- In Ring Section sometimes thin places occurs in yarn which causes yarn breakage in weaving department.

From all these points it is concluded that winding process has its own worth which is not denied in any case.

Tasks of Winding Process

- Extraction of yarn faults
- Manufacturing of large yarn packages i.e. cones, cheese etc having a sufficient yarn length on it.
- To remove spinning faults e.g. thick and thin places.
- To improve the quality of yarn.
- During the winding process, waxes are also applied to the yarn to reduce the abrasion and friction resistance to make the yarn softer.

2.9.1.1 Passage of Yarn in the Winding machine

In a typical winding machine the yarn passes from the supply bobbin through a tensioning device, a slub remove, a tension bracket, over a package lever and finally on to take up package.

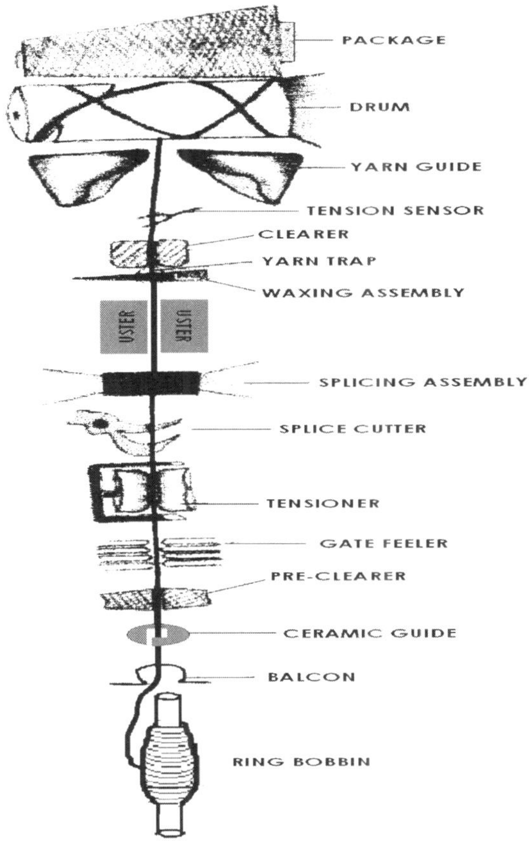

Fig. 2.23 Yarn path through a winding machine

Most modern winding machines are equipped with the thread detectors which detect the presence of thread and if it is broken, it is automatically repaired by knotters or splicers. The bobbins are placed in a magazine and when a bobbin finishes out it can be automatically replaced by a full one.

The tensioners detect whether the yarn passing through them is of the required strength or not. If a yarn has a thin place, a bad piecing or soft ends it will not be able to tolerate the exerted tension and will break. The faulty region is then removed and yarn is knotted or pieced together by auto knotters or splicers. Slub is removed by slub removers, which are of various types.

This is the last section of yarn manufacturing process, after this department yarn goes to the conditioning section for conditioning. After this it goes to the packing department. After packing it goes to the weaving mills for fabric manufacturing. The production of this department is 209 bags/ day.

2.10 Fabric Production

Fabric production or weaving means the making of fabric by yarn, which is made by the interlacement of warp and weft yarn on the loom. These yarns are vertically interlaced. There are different processes involved for making the fabric like warping, sizing, weaving and folding. These processes are dependent on each other. Warp beams are transferred to the sizing process where these beams are sized by different chemicals to give strength to yarn and then, these sized beams are attached in looms for making the fabric. After this, fabric is sent to the folding department where fabric is folded and inspected to check the faults. Faulty fabric is separated from the A, B grade fabric.

Following are the processes through which the fabric is produced step by step.

2.10.1 Warping process

The objective of warping process is to make the weavers beam for the weaving machine because it is impossible that to place the hundred of cones in front of weaving loom to provide the warp threads.

Types of warping processes

Direct warping

When several beams with same beam length are prepared then direct warping is used. In this process the proper winding takes place, the available threads from the creel are wound on the beam. After this all available beams are unwound on the weavers beam.

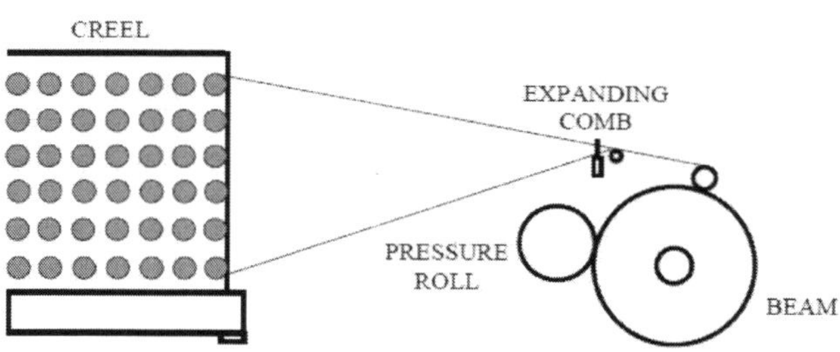

Fig. 2.24 Direct Warping

Sectional warping

In this system several sections are wound in the sequence which parallel to each other on a drum. This type of warping is cost effective for short and striped warp and fancy pattern. Here the warping speed is about 800 m/min while the beaming speed is about 300m/min.

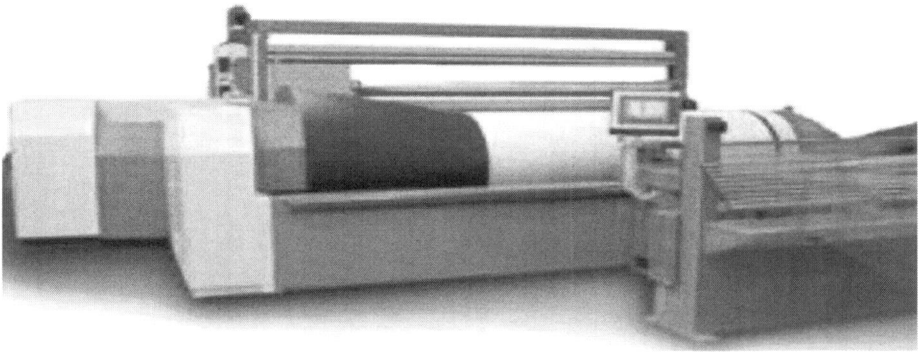

Fig 2.25 Sectional warping

2.10.2 Sizing

Strong, smooth and elastic yarn is the requirement of weaving process. To achieve these properties a polymeric film or coating is applied on the warp yarn

Following are the objective of sizing process

Strengthen the yarn

- ❖ Make yarn surface smoother
- ❖ To reduce the yarn hairiness

The goal of the sizing process is to reduce the warp breakage during weaving. Warp breakage is either caused by low strength yarn or through high tension.

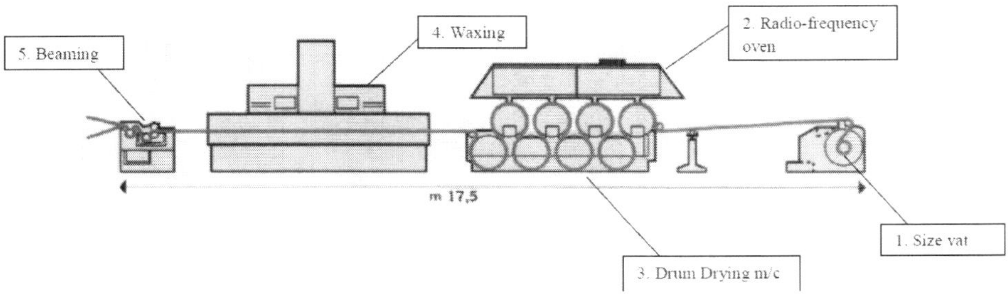

Fig 2.26 Sizing

The important ingredients which are used in the sizing recipe are Adhesives, lubricants, additives, water like PVA, CMC, Starch etc

Following are the parameters which controlled in the size box:

- ❖ Level of size solution
- ❖ Temperature

- Size liquor concentration
- Squeeze roller pressure
- Yarn speed

2.10.3 Drawing In

This process consists of threading through each drop wire, healed wire and reed dent depending on style and size of machine. This operation is carried out manually and now a day's automatic drawing machines are available.

Fig 2.27 Drawing

2.10.4 Tying In or Knotting

When the fabric is mass produced and weaver beam is about to finish then, new weaver beam is placed on the loom having same ends, each end is tied to its corresponding end of loom. This process is carried out through knotting machine

Fig 2.28 Tying

2.10.5 Weaving Process

Weaving is the interlacement of warp and weft yarn. The machine which is used for this process is call loom; both yarns are interlaced at right angle, warp yarn is working in vertical direction while weft yarn is inserting in horizontal direction

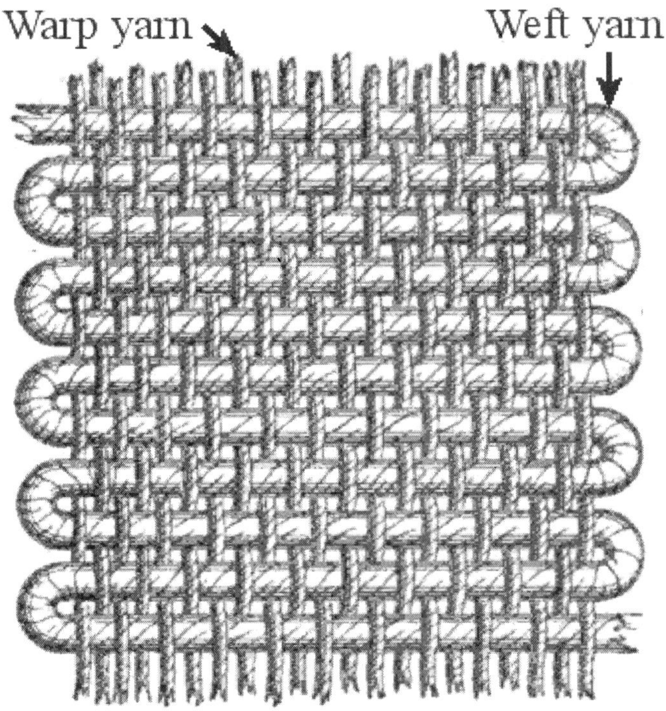

Fig 2.29 Fabric weave

Many types of looms are available for fabric production and are being used in different companies.

2.10.6 Loom

Different types of looms are used in KTM. The details of these machines are given below:

2.10.6.1 Projectile loom

In this loom filling of yarn is done through the small metal device just like a bullet in appearance with a clamp to grip the yarn from one end then, this metal device propelled into the shed. This loom is considered as a high productivity level, having 450 rpm and reliable operational functions.

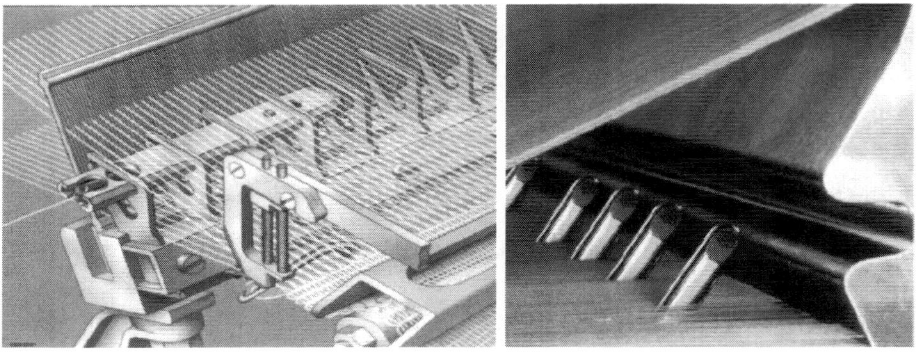

Fig 2.30 Projectile loom

2.10.6.2 Air Jet loom

In Air Jet loom the weft insertion is done through air pressure. This machine has the highest weft insertion performance and it is considered as the most productive in case of all type of fabrics from light to medium weight fabric.

The weaving width normally ranges from 190 to 400 cm and has the capacity of 8 different colors fed at a time. Although, it requires high energy consumption to prepare compressed air and this consumption increase with the increase of loom width and running speed.

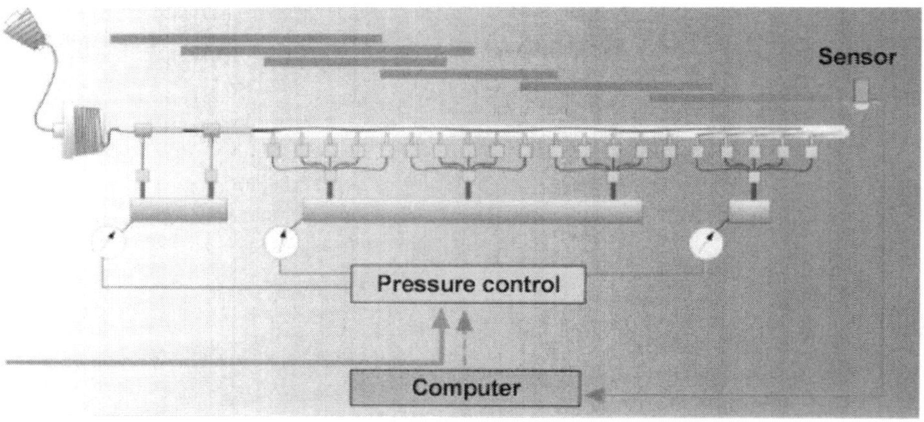

Fig 2.31 Source

2.10.6.3 Water Jet loom

Only few companies produced this type of machines, normally used for light to medium weight fabrics, in water repellent materials and mainly utilize for multi filament synthetic yarns. Water Jet loom are extensively used in East Asia while on other countries it have limited importance. They are categorized as high insertion rate with low energy consumption.

Weft yarn fed from the cone through pressurized water, from the water pump the weft yarn is inserted to the shed at high speed

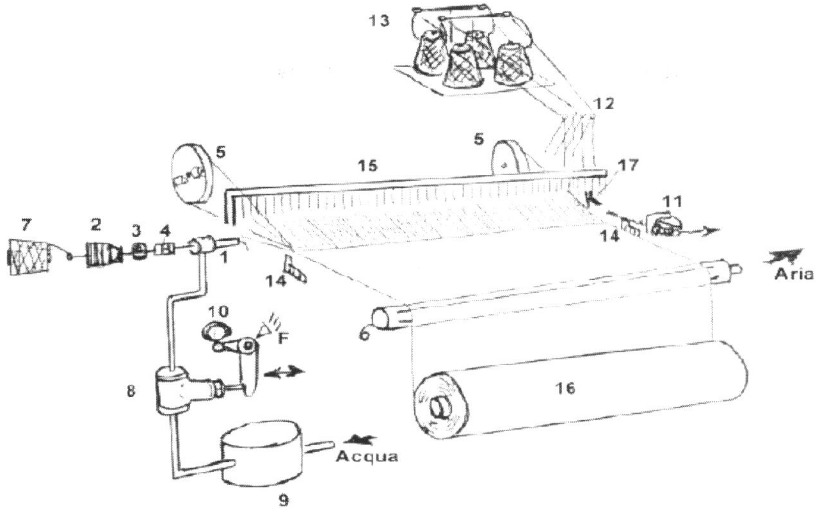

Fig 2.32 Water jet loom

2.10.6.4 Rapier loom

Rapier loom is the most flexible machine in the market; its application covers a wide variety of different fabric styles. Rpm of this machine is about 600-700. During working reed, healed frames and slay has low vibration. The weft yarn is inserted through clamp containing the weft passed through the shed and then comes back to its initial position; this is the example of single

Fig. 2.33 Rapier loom

2.11 Weaving Mechanism

The mechanism of any loom can be classified as described below.

- Primary motions
- Secondary motions
- Auxiliary motions

The primary motions contain the shedding, picking and beat up motions.

Shedding means the open of warp sheet into two layers to make the way for the passage of shuttle containing the filling yarn.

The *Picking motion* varies from loom to loom e.g. in air jet weft picking is done through pressurized air.

Beat up is the process of pushing the last inserted weft yarn to shed fell into cloth by another device called reed.

Secondary motions consist of let off, take up and auxiliary motions.

In take up motion, as the fabric is weaved the first task is to remove it from the weaving area, to achieve this take up motion is used. Take up motion support to wind the cloth on the cloth roller .The fabric removes the cloth at rate that controls the filling density pick/ inch.

In let off motion, warp yarn from the warp beams are releases, as warp is woven into the fabric it remains constant throughout the whole weaving process.

Auxiliary motion consists of warp stop motion, warp protector and weft stop motion. Warp stop motion is used when warp yarn breaks then loom will stop automatically with the help of sensors. Warp protector is used for several warp thread breakages in the happening of shuttle getting trapped in the middle of warp sheet. Weft stop motion is used when weft yarn breaks then through sensor loom stops automatically. .

Chapter 3

The Ring Spinning Process

3.1 Introduction

The ring spinning is the final operation in the formation of the ring spun yarn. The basic purpose of the ring spinning frame is to attenuate the roving until the required fineness of the yarn is achieved. The ring spinning machine was first invented in 1828 in America. Since then a lot of modern modifications have been carried out but the basic principle of operation and the basic machine design remains the same. Even with the introduction to new sophisticated spinning systems, the ring spinning system is widely used and is the most popular form of spinning system. The reason behind is the ability of this system to produce very fine yarns which is not possible to obtain at the moment with any other form of spinning systems. Only the rotor spinning system provides good competition to the ring spinning system which again cannot produce fine yarns of the order produced by the ring spinning frame. A staple yarn may be defined as a continuous collection of fibres held together by a binding medium such as twist.

Since before 1900, ring spinning has been and still is the method used to produce most of the world's yarn. For this reason, we will review the basic principles of ring spinning before proceeding with a description of open-end spinning.

Ring Spinning

The basic requirements and functions of a ring-spinning frame are:

a) fibre supply,

b) drafting,

c) twisting, and

d) package winding.

The fibre supply, usually in the form of roving and having a hank* range from approximately 0.40 to 5.0, is drawn into the drafting system by the rotation of the back rolls. In the drafting zone, the weight per unit length of the input roving is reduced because the surface speed of the front drafting rolls is greater than that of the back rolls. Normal ring frame drafts range up to about 30.

The front rolls deliver a continuous cohesive stream of fibres, which must be twisted immediately into yarn. This transformation is accomplished by the interactions of the spindle, ring and traveller. The rotation of the spindle causes twist to be inserted into the stream of fibres delivered by the front rolls. By passing the yarn under the traveller, winding is accomplished.

The spindle serves three functions: first, it provides a location to wind a package; secondly, by rotating the yarn package, the spindle causes twist to be inserted into the strand of yarn being formed at the nip of the front rolls; thirdly, the rotation of the package causes the yarn to pull the traveller around the ring, providing a method of not only transmitting twist derived from the spindle, but also a guide to change the direction of yarn travel so that it approaches the yarn package tangentially, and thus can be wound onto the package.

The initial and maintenance cost of the ring spinning is quite high. The ring frame contributes about 60% towards the final cost of production of the yarn. So every effort is being put in to make modify the ring spinning machine to give more productivity at a less cost.

3.2 Functions of the Ring Spinning Frame

The basic function of the ring spinning frame is drafting. Drafting is carried out to such an extent so as to achieve desired fineness of the yarn. The drafting is carried out by using a 3 over 3 roller drafting arrangement with double apron support. At the ring spinning frame a draft of 15 to 40 is given (sometimes also up to 50).

The attenuated yarn formed by drafting is weak and lack cohesion. In order to give strength to the final yarn, twist is inserted. Twist is inserted at the ring spinning frame by using the popular traveller method. The amount of twist inserted in the yarn varies with the count of the yarn. To make finer yarns, greater TPI is given and vice versa. The value of Twist Multiplier (TM) at the ring spinning frame generally ranges from 3 to 5.

The final yarn produced after drafting and twisting is wound on special ring bobbins also called as cops. The built of the package kept is such which is suitable for storage, transportation and further processing. The schematic diagram of the ring frame is given in Figure 3.1.

The ring spinning frames are double sided or double delivery machines having up to 500 delivery points or spindles on each side. Hence a single machine can have a capacity of up to 1000 spindles.

A typical ring spinning frame can be divided into the following four zones:

- Creel Zone , Drafting Zone
- Twisting Zone, Package Winding Zone

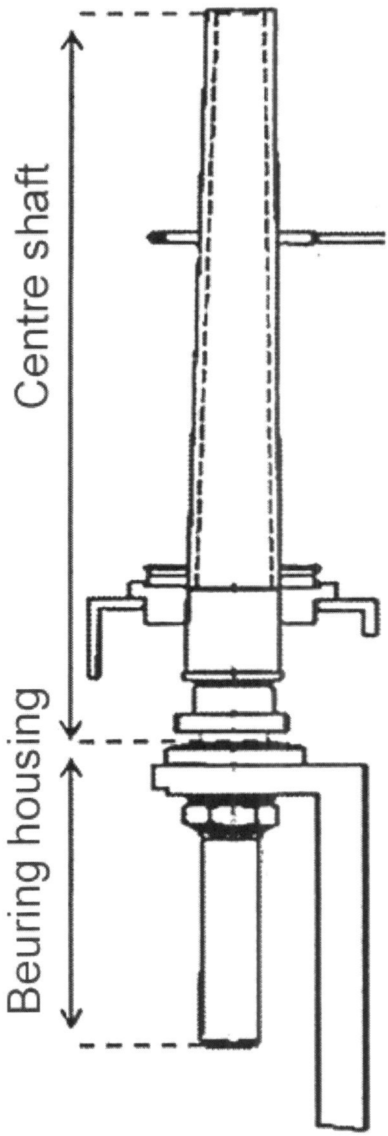

Figure 3.1 The typical diagrammatic representation of a ring frame and its spindles

3.2.1 Creel Zone

The creel of the ring spinning frame is a simple device that holds the roving. It is very important that the roving should unwind properly and evenly to avoid any false drafts or roving breakages. To facilitate proper unwinding the roving bobbins are held vertically by inserting the roving bobbin on the bobbin holder placed at the upper portion of the frame. A typical bobbin holder is shown in the figure 3.2 below:

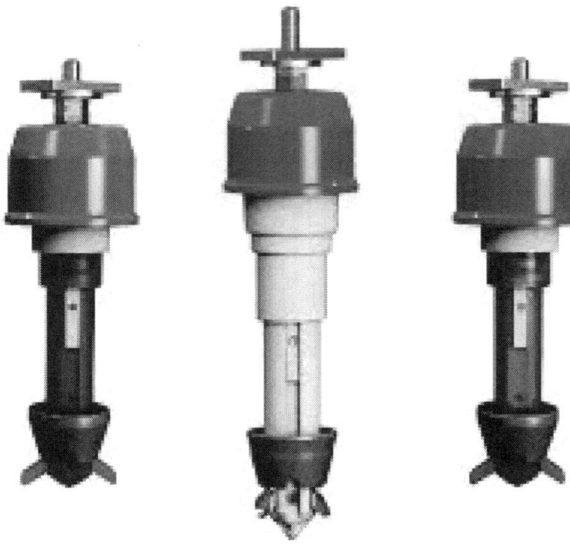

Figure 3.2 A typical bobbin holder

The upper end of the bobbin holder is suspended by a bearing. The roving bobbin is pushed upward on to the bobbin holder. Since the ring of the bobbin holder is attached to the retainer device so as the bobbin is pushed upward, the ring will also move upward causing the retainer device to grab the inner portion of the roving bobbin tube. As the roving is pulled forward by the drafting rollers, the roving bobbin rotates and unwinding takes place. To avoid excessive rotation to the bobbin, a brake arm is used to lightly press the roving bobbin to restrict its motion. Alternately this can also be achieved by using internal brakes in the ball bearings.

3.2.2 Drafting Zone

Since the at the ring spinning frame greatest value of draft is given, the drafting zone of the ring frame is the most important part of the machine that directly influences the evenness and strength of the yarn. So it is important to have a precision control on the fibres in the strand which is being drafted out to suppress the drafting waves.

Pendulum Arm for Weighting

All the modern ring spinning frame makes use of a 3 over 3 roller drafting with both thetop and bottom middle rollers covered with aprons. The bottom rollers are steel fluted and are driven positively. The top rollers are covered with synthetic rubber coating and are driven by the frictional contact with the bottom rollers and are hence pressed down on the bottom rollers with a considerable force. For this purpose the top rollers are weighted by using spring, pneumatic or magnetic weighting with the help of a pendulum arm. Please refer figure 3.3 below.

A total draft of 15 to 40 and in some cases even up to 50 can be given at the ring spinning frame. The distribution of draft in the back and front drafting zone is very important as a proper draft distribution helps to reduce the drafting wave problem. Practically the break drafts should be given in the range of 1.1 to 1.5 and the remaining draft is given in the main drafting zone. High degree of drafting in the front zone will tend to create drafting waves.

To avoid this middle pair of drafting rollers is covered with aprons. Aprons give support to the fibres throughout the roller spacing between middle to front rollers not letting even the shortest fibre to float. Apron systems are not perfect but they help to drastically reduce drafting waves. Based upon the type of the bottom apron used, two types of apron systems are used commercially on the ring spinning frame:

- Short Bottom Apron System
- Long Bottom Apron System

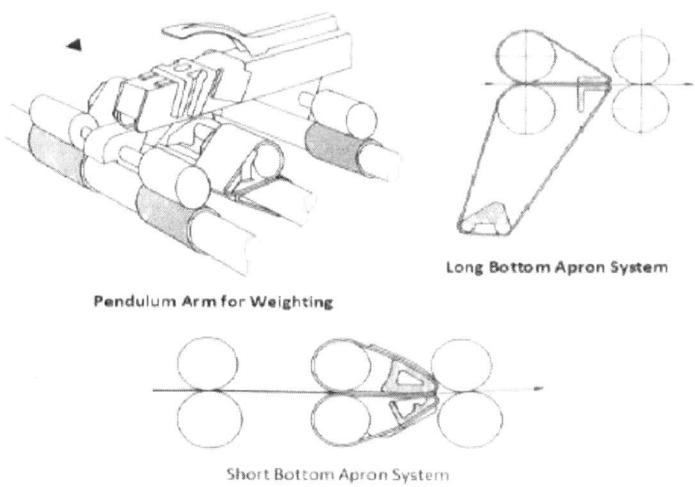

Figure 3.3 Drafting systems for ring frame

The short bottom apron system was used in old machinery. Although the short aprons require a simpler design of the ring frame and also have the advantage that they can be brought more close to the nip of the front rollers, however they are difficult to replace when damaged and can also cause choking of fibre mass. To avoid these limitations the modern spinning machines make use of long bottom aprons.

Using Trumpet Guides

The strand of roving has few numbers of fibres in a cross-section and upon drafting even fewer number of fibres will remain in a yarn. These fibres have hardly any cohesion between themselves. Special fibre guiding elements are therefore essential to carry out drafting without breaking the strand. For this purpose specially designed trumpet guides are used in the drafting zone to help keep the fibre mass together and to avoid tearing apart of the strand. Three trumpet guides or condensers are used one just before the back pair of rollers, second in between the back and middle rollers and the last just before the front pair of rollers. These guides are mounted on reciprocating bar that oscillates slowly so that the wear of the rollers may spread out gradually throughout the entire roller width.

Overhang of Front and Middle Top Roller

As the twist is given to the yarn with the help of a traveller, the twist must run back as close as possible to the nip line of the front rollers. But it never penetrates completely to the nip because at the nip the fibres are diverted inwards and are made to wrap around the surface of the roller.

So at the exit point of the front pair of rollers there is always a small triangular mass of fibres which remain without twist, this is called as the spinning triangle. This triangle bundle of fibres will remain without twist during twisting and create a weak point in the yarn and hence breakage rate of the yarn during spinning increases. The effect of spinning triangle is enhanced at the ring spinning frame where greatest draft along with greatest twist is applied. Please refer figure 3.4.

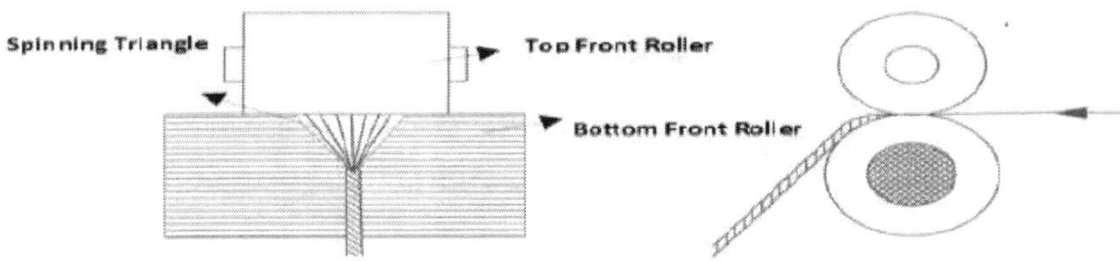

Figure 3.4 Diagrammatic representation of the spinning triangle

To further enhance the flow of the fibre strand through the drafting arrangement, the drafting roller instead of being placed in a straight horizontal line as depicted from the above figure, they are arranged at certain inclination. Please refer figure 3.5.

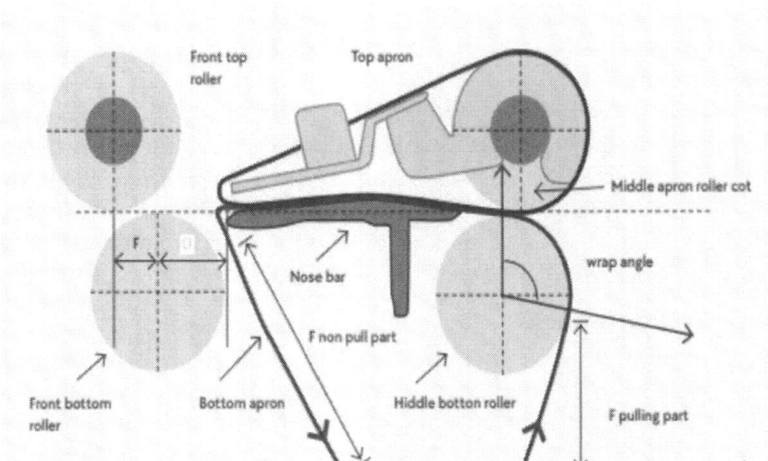

Figure 3.5 The schematic arrangement of the drafting zones on the ring frame

3.2.3 Twisting Zone

The yarn leaving the front pair of drafting rollers is threaded through a guide also called as the lappet guide that is placed directly over the spindle axis. The yarn then passes through a traveller on to the yarn package (cop). The traveller is mounted on the ring encircling the spindle. The cop is mounted on the spindle and rotates with the spindle. When the cop rotates the tension on the yarn pulls the traveller to rotate around the ring. Due to the friction created between the traveller and the ring and also due to the air drag created by the balloon formation at the winding section, the rotary and surface speed of the traveller is slightly less than that of the spindle and is given by:

RPM of Traveller =RPM of Spindle – Delivery Speed / π x Package Dia

So in one rotation of the traveller one twist is inserted in the yarn. Since the difference between traveller and spindle rotary speed is very small so this difference is ignored. Thus for calculating twist rotary speed of the spindle is considered, mathematically:

T.P.I = Rotary Speed of Spindle (rpm) / Delivery Speed (inches/min)

The ring/traveller arrangement and the spindle are also very important elements of the ring spinning frame because it not only helps to insert twist in the yarn but also helps in the winding of the package.

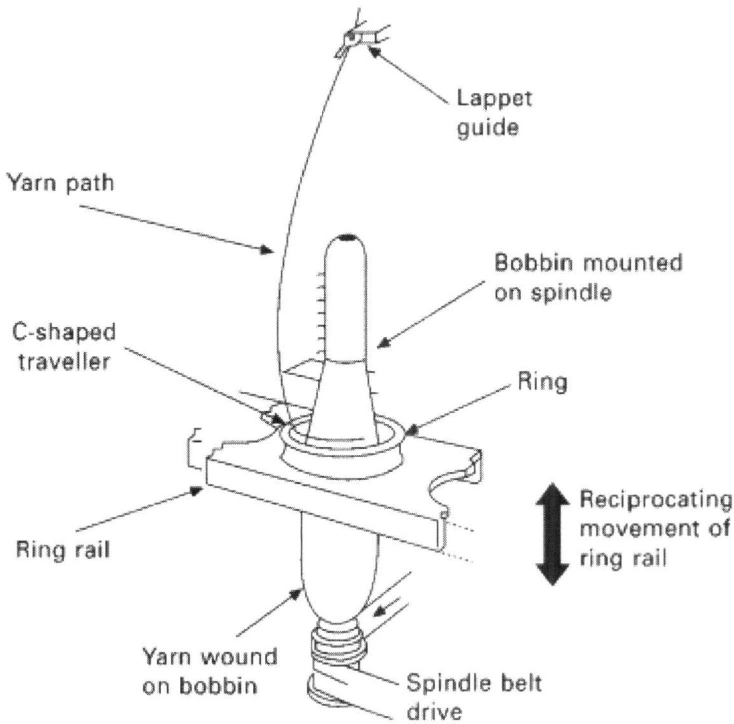

Figure 3.6 The schematic arrangement of the spinning elements of the ring frame

So, in the ring spinning drafting arrangement the front top roller is deliberately kept 2-4 mm forward relative to the bottom front roller while the middle top roller is kept 2-4 mm backward relative to the middle bottom roller. This type of arrangement gives smooth running of the top roller and helps to reduce spinning triangle and ultimately the breakage rate during the spinning process is reduced.

The ring/traveller arrangement and the spindle are also very important elements of the ring spinning frame because it not only helps to insert twist in the yarn but also helps in the winding of the package. Please refer to figure 3.6.

The Ring

The rings are supported on the raise-able and lower-able ring rail. The ring should have a tough, hard and smooth surface. For this purpose flame hardened steel, nitrided steel, carbo-nitrided steel and chrome steel is used. The hardness of the traveller should be less than that of the ring so that wear mainly occurs on the traveller which is cheaper and easier to replace.

The rings can either be classified as:

- Lubricated Rings
- Un-lubricated Rings

Lubricated ring is used for woollen and worsted spinning where external lubrication on its surface is applied. For cotton and other short staple spinning only un-lubricated rings are used. This is due to the fact that that all natural short staple fibres are made up of cellulose and has small quantity of natural waxes in it. As the yarn moves swiftly with the traveller, a lubricated layer of waxes and cellulose is generated due to the abrasion of the fibres. This lubrication is beneficial for decreasing the friction between the traveller and the ring and ensures a smooth continuous running. The amount of lubrication generated depends mainly upon the count of the yarn spun and quality of raw material being used. Coarser yarns generally produce more lubrication. On the basis of working sides of the ring, rings can also be classified as:

- Single Sided Rings
- Double Sided Rings.

Double sided rings have the advantage of having two working sides. When one side gets abraded the ring can be turned over so that the other side can be used. Whereas when single sided rings get damaged, they have to be replaced by new ones. However double sided rings are not used anymore because when it is time for the ring to be turned over, the other side of the ring will become unstable due to corrosion and rusting and cannot be used.

The Traveller

The traveller is an important element of the ring frame as it imparts twist on the yarn and also helps in winding the yarn on the cop. The traveller does not have a drive of its own but is dragged along with the yarn as it gets wound on the cop due to the rotating spindle. As the traveller moves at a high speed on the ring it creates lot of friction that generates significant amount of heat. So in order to cope up with such frictional forces the material used for traveler should exhibit the following properties:

- Generate as little heat as possible.
- Should dissipate heat quickly to the ring and the air.
- Should be elastic so that it might not break.
- Should have high wear resistance.
- Should have less hardness than the ring.

To meet these requirements, the traveller is made exclusively of steel having its surface either electroplated with nickel and silver or treated with chemicals to increase wear and reduce friction.

Different types and varieties of yarns cannot be spun using a single type of traveller. Therefore travellers come in many different varieties. The differences among travellers are found in shape, mass, wire profile and yarn clearance (bow height). Please refer figure 3.7.

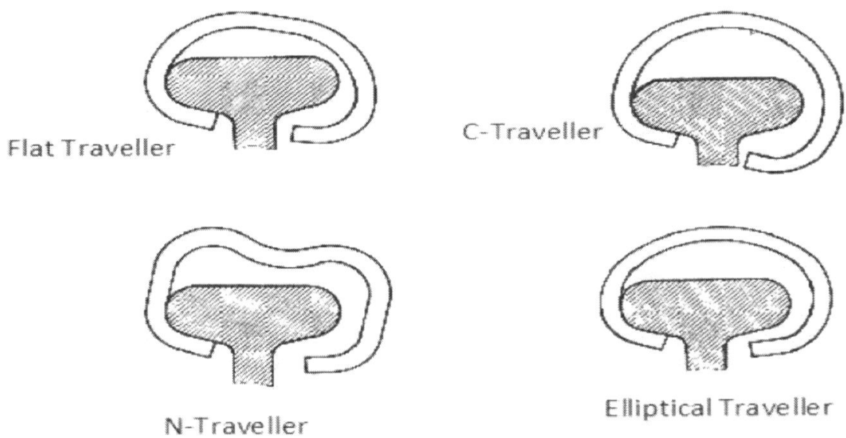

Figure 3.7 Different types of travellers

The traveller is shaped according to the ring so that single point of contact with greatest possible surface area with the ring can be achieved. The height or bow of the traveller should be as low as possible to keep the centre of gravity low and to ensure smooth running. However the bow should be not so small where the yarn starts getting abraded which can damage the yarn hairiness.

Another important parameter of the traveller is its mass. The mass of the traveller directly influences the frictional forces between itself and the ring. The mass of the traveller is selected according to the yarn count being spun. If the traveller is too light for a specific yarn count then tension created on the yarn will be low resulting in a loose softly wound cop and also the size of the balloon increases. On the other hand, if the traveller is too heavy then it causes excessive tension on the yarn and cause frequent yarn breakages. For convenience instead of the traveller weight they are standardized by its number. The following table (Table 3.1) shows various traveller numbers suitable for different yarn counts:

Table 3.1 Traveller number for different cotton counts

Yarn Count	Traveller Number For Cotton	Traveller Number For Synthetics and Blends
85	16	22.4
75	20	28
65	23.6	31.5
60	26	35.5
50	31.5	40
42	35.5	45
35	45	56
30	56	71
24	71	90
20	90	100
16	100	112

Limitations of Traveller/Ring Arrangement

The greatest limitation faced by ring spinning frame is its incapability to further increase its productivity and speed. This is mainly due to the use of traveller ring arrangement. At high speeds the traveller generates significant amount of heat due to friction with the ring. If the speed of the ring frame is to further increase, owing to a very small mass and size of the traveller, dissipation of heat generated by the traveller is not possible. Thus the operating speed of the traveller and the ultimately the entire machine is limited. Currently the spindle of the ring frame is capable of running up to 30,000 rpm but the use of traveller permits the speed only to be raised up to around 22,000 rpm.

Another way of increasing the speed of the ring frame is by using ring with smaller diameter so that spindle speed can be further increased without increasing the frictional forces on the traveller/ring. However by doing so smaller packages will be formed and they will have to be doffed more frequently. This can decrease the productivity of the machine. However this problem can be rectified by using automatic doffing devices.

To put into a nutshell, there is a very small room as far the speed advancement in the ring spinning frame is concerned. However measures like automatic doffing, use of precision instruments and precision components (especially ring and traveller), increase in the number of spindles per machine to reduce cost per spindle and automatic linkage of ring frame with the winding machines will help the ring spinning frame to be economically viable to use for years to come. Limitations: Mechanically speaking, the production of a ring frame is limited by excesses in one or more of the following:

a) spindle speed,

b) front roll speed,

c) traveler speed,

d) frame down-time for doffing.

Economical limitations on ring spinning are related to power consumption and package size. Power requirements to rotate the package are greater than that needed to insert twist only. The package size is limited since it must be enclosed in the yarn balloon. Finally, for very coarse yarn counts, frame down-time for doffing must be considered since it increases as coarser yarns are spun.

3.2.4 Package Winding Zone

The yarn after being twisted by the traveller has to be properly wound on the entire length of the package called as the cop. In order for the winding to take place there should be a difference in the surface speed between the spindle and the traveller. The traveller has no drive of its own instead it is carried along with the yarn which is being wound on the cop after passing through the traveller. The frictional forces between the spindle and the traveller and also the air drag because of the balloon formation between the lappet guide and the spindle makes the surface speed of the traveller less than that of the spindle. This automatically fulfils the requirement of winding. The traveller and the spindle cooperate with each other to perform winding.

In order for the yarn to be properly wound on the entire length of the cylindrical ring bobbin (cop), the ring is raised and lowered which is mounted on a longitudinal ring rail.

The Spindle

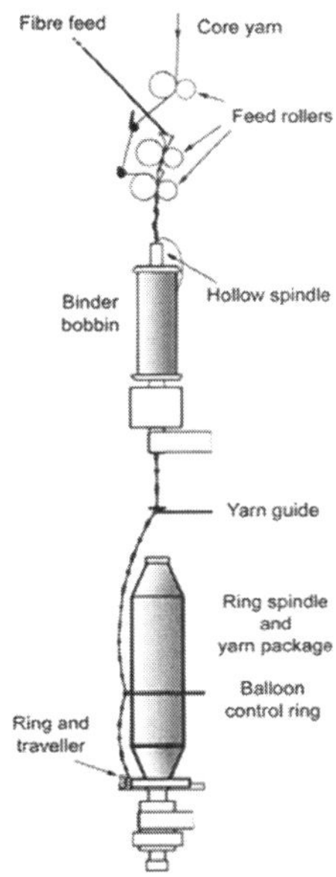

Figure 3.8 Diagrammatic representation of a ring frame spindle

Spindle is a very important component of the ring frame, it not only supports the cop and helps in winding of the yarn on it but also gives motion to the traveller. The spindle and the traveler therefore work in close cooperation with each other to perform both twisting and winding. A spindle is a metallic shaft that holds the yarn package or cop. The spindle is divided into two main parts namely the upper part and the bolster. Please refer to figure 3.8

The upper part is made of aluminium alloy and is slightly tapered to firmly hold the package. The bolster or the lower part is secured in a bearing to rotate freely and is also connected to the ring rail by means of a nut. The drive to the spindle can either be given with the help of tapes or belts in group form or in individual form where one motor is supplied per spindle.

3.3 Package Built of Ring Spinning Frame

A typical form of a package produced by a ring spinning frame is called as the cop. The package is produced on a tube of paper, card board or plastic material. About 10 mm of the tube is left free of the yarn on both upper and lower ends. The tube has a slight taper corresponding to the taper of the spindle so that the package can be firmly secured on the spindle.

Package Tube

The structure of the cop can be divided into three distinct parts as shown in figure 3.9:

- ❖ The Upper Conical Tip.
- ❖ The Middle Cylindrical Part.
- ❖ The Lower Curved Base.

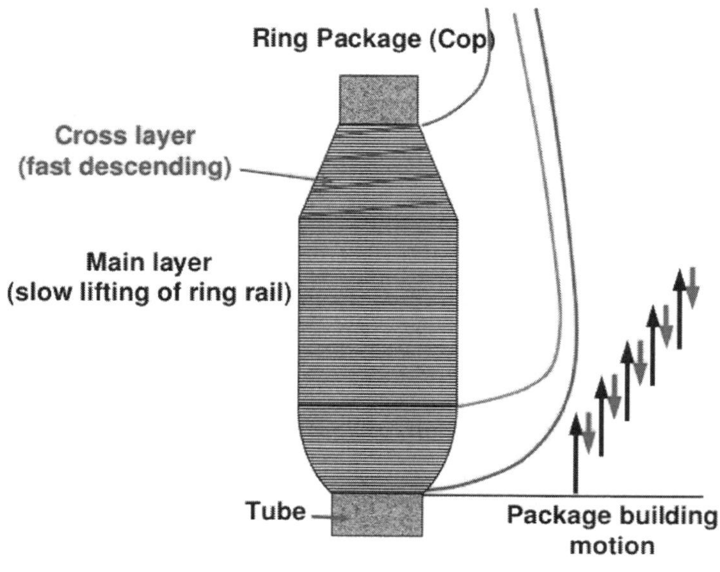

Figure 3.9 The schematic structural diagram of a ring cop

The cop has two distinct layers of the yarn wound on its surface i.e. the main winding layers and the cross winding layers. The main winding layers are close to each other and are densely wound whereas the cross winding layers are openly laid and are loosely wound. The cross winding layers separates the main winding layers so that the high tension of the yarn may not cause the successive layers to be pressed causing problems in unwinding. The package built of the cop as compared to that of the roving bobbin is much more complex which also requires complex mechanical arrangements; however it is optimal for unwinding at a very high speed on the winding machines.

The Builder Motion

To obtain the above mentioned built of the cop, complex mechanical arrangements called as the builder motion is provided at the ring frame. The builder motion achieves the structure of the cop by performing the following tasks:

The ring rail on which the ring is mounted is moved up and down by the builder motion in a special manner. The ring rail while moving up is moved slowly but with an increasing speed and is moved down quite faster but with a decreasing speed. During the slow upward movement of the ring rail, a dense main winding layer is laid and during the fast downward movement, a loose cross winding layer is laid. This gives the ratio between the length of the yarn in the main and cross windings to be 2:1. This difference in yarn length in the upper and middle part creates a taper or conical shape of the upper part. The stroke or the distance by which the ring rail moves up and down is less than the total winding height (lift) of the tube.

The stroke of the ring rail is deliberately kept less than package tube height so that the ring rail can be moved upward by a small distance from the previous point of lay of the yarn after one complete layer of the yarn has been laid. This helps in creating the lower curved base of the package.

In older machinery, instead of moving the point of lay of the ring rail upwards after every layer has been laid, the spindle is made to move downward. This is carried out by mounting the spindle on the spindle rail. However in modern machinery this arrangement is no longer used and only the ring rail is made to move up and down and also its point of lay is continuously lifted upwards by a small amount. This somehow reduces the complexity of the builder motion.

The tasks of the builder motion to move the ring rail up and down is carried out by using special heart shaped eccentrics or cams in its drive given to the ring rail. These eccentrics in its rotary motion give an up and down motion to the ring rail. The rise in the point of lay of the ring originates by using a ratchet wheel that after every layer of the yarn makes the pint of the lay of the ring rail to shift upwards.

3.4 Use of Balloon Control Rings (BCR)

To achieve greater production speeds with minimum friction produced by the traveller, the spindle and the rings used on modern ring frames are comparatively small. This small ring size only permits a very limited length of yarn to be wound on the bobbin and the doffing would have to be performed more frequently and hence the productivity would suffer. On modern machines this problem has been overcome by using longer length of spindles and bobbins so that even with less package diameter more yarn length can be wound. This reduces the amount of doffing and productivity can be increased in this way.

The use of long spindles will mean that the distance between the lappet guide and the ring would be quite high resulting in a formation of a bigger balloon. The greater balloon size has following disadvantages:

To facilitate a larger balloon during winding, more spacing will be required between the adjacent spindles and hence the size of the machine has to be increased.

The large balloon causes the air drag on the yarn to increase and due to which the yarn can be broken due to balloon collapse.

In order to avoid these disadvantageous effects, balloon control rings are used. Each balloon control ring divides the balloon into two smaller sub balloons which are more stable and create less air drag on the yarn. The use of balloon control rings and separator is illustrated below in figure 3.10:

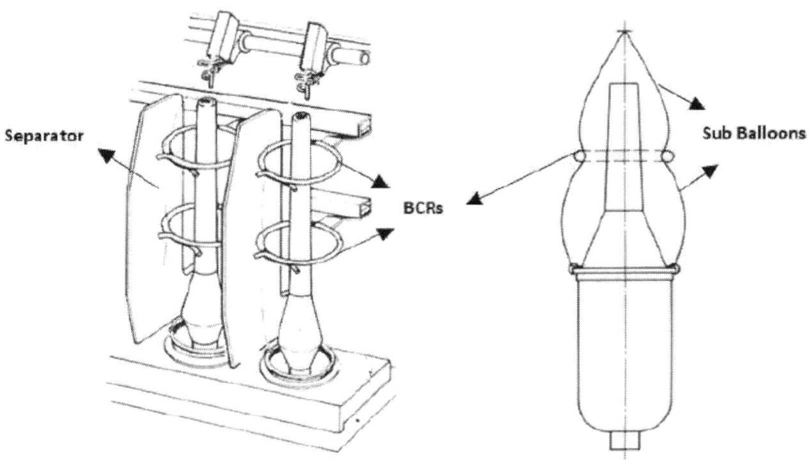

Figure 3.10 Schematic view of balloon control ring and separator

The balloon control rings enable the spinning operation with long spindles at high speeds. However the use of such rings cause the surface of the yarn to become rough and abraded as the yarn is rubbed against the surface of these rings. This can also create more fly in the spinning department.

3.5 Use of Separators

Most of the end breakages during spinning occur because of the weak spots created by the spinning triangle. When the thread breaks the free broken end of the thread lashes around the spindle. If a protective plates called as separators are not provided to separate the adjacent spindles, the broken end of the thread can get entangled into the neighbouring spindles causing more threads to break. So a single end breakage can cause a chain reaction of yarn breakages in all the spindles placed in one line. To prevent this separator plates made up of aluminium or plastic is placed in between every adjacent spindle.

3.6 Compact Spinning

Introduction: The developments in spinning systems had been happening and hence, the industry started with mule spinning and then moved on to cap spinning, flyer spinning and then ring spinning up to today's latest spinning systems like rotor spinning, air jet spinning and so on. Now, there are many practically proven and commercially viable ways to produce yarn. However, the versatility of ring spinning has created its own firm root in today's spinning industry. Currently, the compact spinning technology is one of the major developments in the history of ring spinning, as the improved properties of compact yarn are much better when compared to the conventional ring yarn.

Compact spinning is essentially a ring spinning with additional feature to eliminate or reduce yarn hairiness in the so called condensing zone. This is done to improve surface integrity and increase yarn strength. The spinning triangle is reduced or eliminated to condensing the fibres in a narrow path. The close and parallel device is closely situated before the twist imparting.

There are two basis systems, air flow or mechanical one. Compact spinning offers near perfect structure by applying air suction or magnetic system for condensing the fibres before twisting, thereby eliminating of the spinning triangle. Compact spinning has been shown to significantly improve yarn tensile properties and reduction in yarn hairiness. Both characteristics are crucial for performance in downstream manufacturing operations. Compact spinning was introduced as a trial system in 1995.

Basic Problems Associated with Conventional Ring Frame:

Basically, the restriction of twist flow to the front roller nip, because of spinning geometry, causes the formation of spinning triangle. The width of fibre strand fed into the spinning nip is always more than width of spinning triangle. The width of spinning triangle depends upon spinning tension and is directly related with spinning tension.

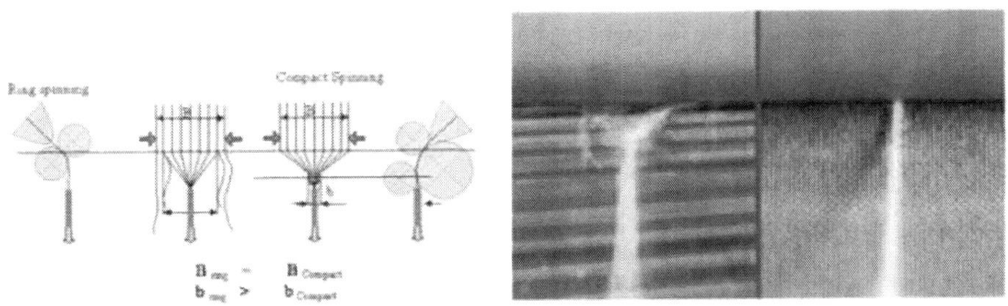

Figure 3.11 The schematic concept of spinning triangle

The fibres at selvedge of fibre strand, delivered by front roller may not get fully integrated into yarn body or they may get lost through pneumafil suction tube (PN waste), because in the spinning triangle, the selvedge fibres are always at very high tension than inner ones and due to improper integration of selvedge fibres into yarn, the hairiness of the yarn rises up.

Also the spinning triangle is the major weak place in yarn formation zone and therefore most of the end breaks in ring frame are occurring in spinning triangle and this is largely influencing the end breakage rate during spinning and hence the machine efficiency. Therefore, for better control of spinning, it becomes necessary to have control over formation of spinning triangle and this is made possible by compact spinning.

3.6.1 Principle of Compact Spinning

The compact spinning is a process where fibre strand drawn by drafting system is condensed / compacted in width before twist is inserted to it.

Following methods are used by some of the leading machine manufacturers to condense/ compact the fibre strand at the spinning triangle.

1. Aerodynamic condensing.
2. Mechanical condensing.
3. Magnetic condensing.

The most important requirement for perfectly spun compact yarn is a complete parallel arrangement of fibres held in close position before twist is imparted. By keeping this core concept into consideration different machine manufacturers have developed different methods of compact spinning.

3.6.2 COM4spin System of Rieter:

In this system, the condensing is done by aerodynamic force. The delivery bottom roller of drafting system is replaced by perforated drum. The fixed suction system under the perforated drum creates a partial vacuum, which generates air current on drum. The air guide element plays important role in condensing process, as air is guided by air guide element & this air current helps to condense/compact the fibre strand. This condensing of fibre strand considerably reduces the width of spinning triangle and hence problems associated with spinning triangle. The yarn produced on this system is named by Rieter as COM4 yarn. Where COM stands for comfort and 4 indicates the four basic advantages of COM4 yarn mentioned below.

1. Reduced hairiness.
2. Maximum strength and elongation.
3. Minimum environmental impact.
4. Unequalled wearing comfort.

The COM4 value can be calculated, COM4 value=100000 / [yarn twist tpm*yarn hairiness H] This value indicates the improvement in COM4 yarn quality.

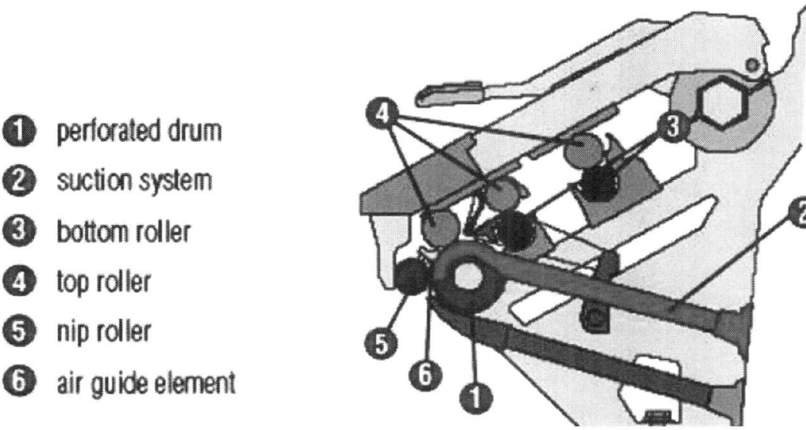

Figure 3.12 Rieter compacting system

The Rieter compact spinning solution is based on aerodynamic parallelization and condensation after the main draft zone. At the heart of this technology is the perforated drum through which suction is obtained to create air currents to condense the fibres coming out of the main draft zone. The main features of this technology are the perforated drum, the suction system, and the air guide element. The setup of the system is shown in Figure 3.12. After the compacting has been done, the fibre strand needs to be twisted. Hence the spinning triangle is formed, which results in deterioration of the orientation of the fibres leading to hairiness, loss of fibres due to fly generation, etc. Therefore, another nip is given between the Nip roller (3.13) and the perforated drum, which doesn't allow the twist to travel up to the compacting zone reducing the length of the spinning triangle, and thus leading to reduction in the abovementioned occurrences. Also due to compacting and condensation the base of the spinning triangle b (Figure 3.14) reduces when compared to normal ring spinning. This technology is also expensive, due to the fact that suction has to be provided to each individual drum.

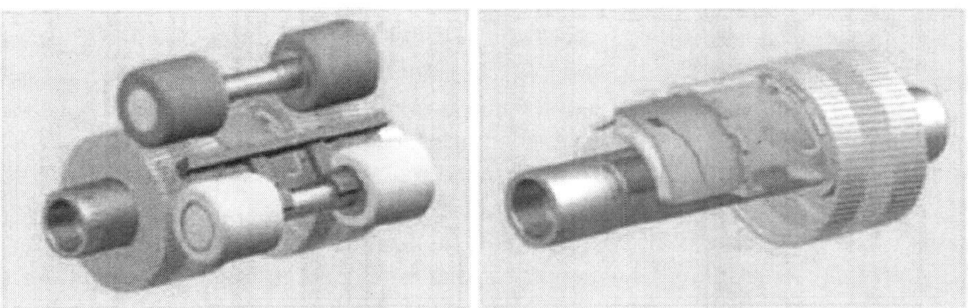

Figure 3.13 : Profile of top roller and perforated drum

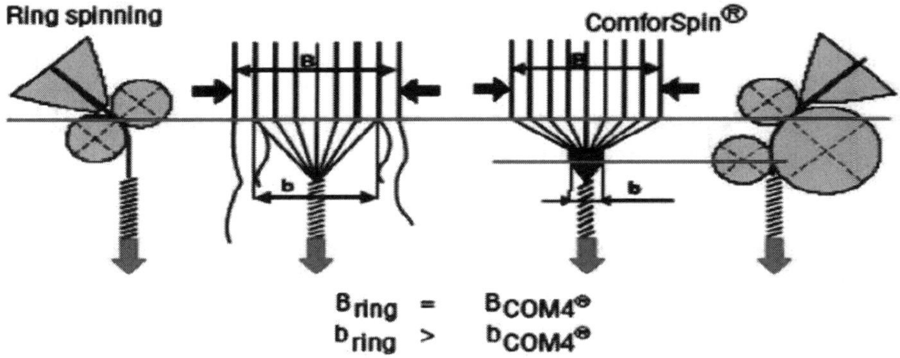

Figure 3.14 Comparison of ring and compact spinning triangles

The Com4 system was conceived by Dr. Ernst Fehrer, the founder of friction spinning. The condensing system of Rieter consists of a perforated drum, just situated after the double apron arrangement and works simultaneously as a delivery roller of the draft system. On the perforated drum, two pressure are situated. The first roller acts as nipping point of the draft system, while the second roller works as twist stop to prevent twist escaping to the condensing zone. The fibre bundling occurs through the suction zone, which is found inside the perforated roller, and in the region between the two pressure rollers. Under this air suction, fibres merging from the delivery nip of the drafting unit are held against the drum surface and moved at the same circumferential speed as the drum surface. In ATME 2000 Rieter South Carolina introduced an innovation in the air guide providing more fibre compactness both against the drum and in lateral direction. This innovation is known as K44-C.om4.

3.6.3 Suessen Elite Compact Spinning System:

Suessen compact system consists of a tubular profile, subjected to a negative pressure and closely embraced a lattice apron. The delivery top roller fitted with rubber cots, presses the lattice apron against the hollow profile and drives the apron, at the same time forming the delivery nipping line. The tubular profile has a small slot in the direction of the fibre flow, which commences at the immediate vicinity of the front roll nipping line in the region of the delivery nipping line. This creates an air current through the lattice apron via the slot towards the inside of the profile tube. The air current seizes the fibres after they leave the front roller nipping line and condenses the fibre strand, which is conveyed by the lattice apron over a curved path and transported to the delivery nipping line. As the slot, being under negative pressure, reaches right up to the delivery nipping line, the fibre assembly remains totally compacted. This results in a substantial disappearance of the spinning triangle. In the Elite system of suessen, the condensing zone following the front drafting roller and it consists of profile tube with suction slot which is placed at certain angle to the flow of fibres, perforated lattice apron and delivery top roller. The lattice apron is driven by delivery top roller and which is driven by gear drive from front top roller of drafting system.

The suction under the tube creates the air currents through slot and lattice apron which are responsible for condensing and perfect parallelization of fibre strand. The suction air pressure, size of slot, speed of apron, number of holds in the lattice apron has an influence on condensing action. Please refer figure next page.

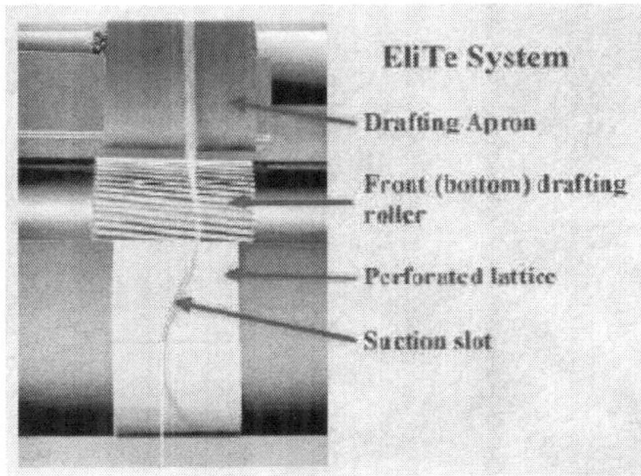

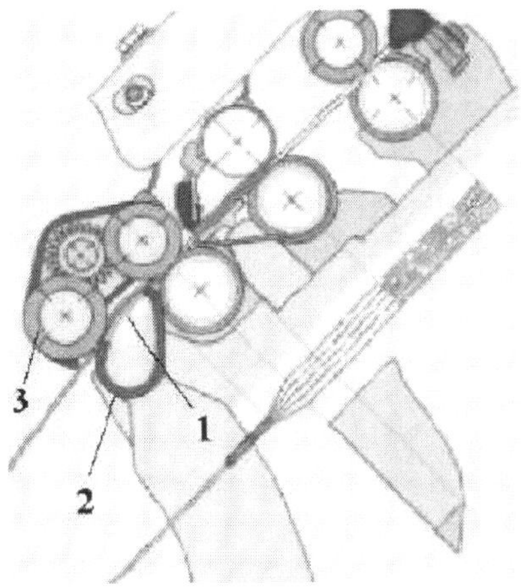

1. Profile tube 2. Inclined lattice apron
3. Delivery Top Roller

Profiled tube with inclined lattice apron

Sussen Elite Compacting System

Figure 3.15 Sussen Elite compacting system

3.6.4 CompACT3 System of Zinser:

The compACT3 system of zinser works on aerodynamic compacting principle. The condensing zone following the conventional 3/3 roller, double apron drafting system and it consists of perforated apron with special vacuum element for correct air current and drive for the perforated apron. The special vacuum element generates the air current underneath the apron. The special arrangement of circular and elliptical perforations on apron insures the better compacting effect. The fibre properties largely influencing the condensing process e.g. stiff fibres behaves in different manner than a flexible ones. The fibre tension in compacting zone has influence on yarn quality. Therefore for cotton compact spinning process, 0-4% overfeeding is required as per raw material.

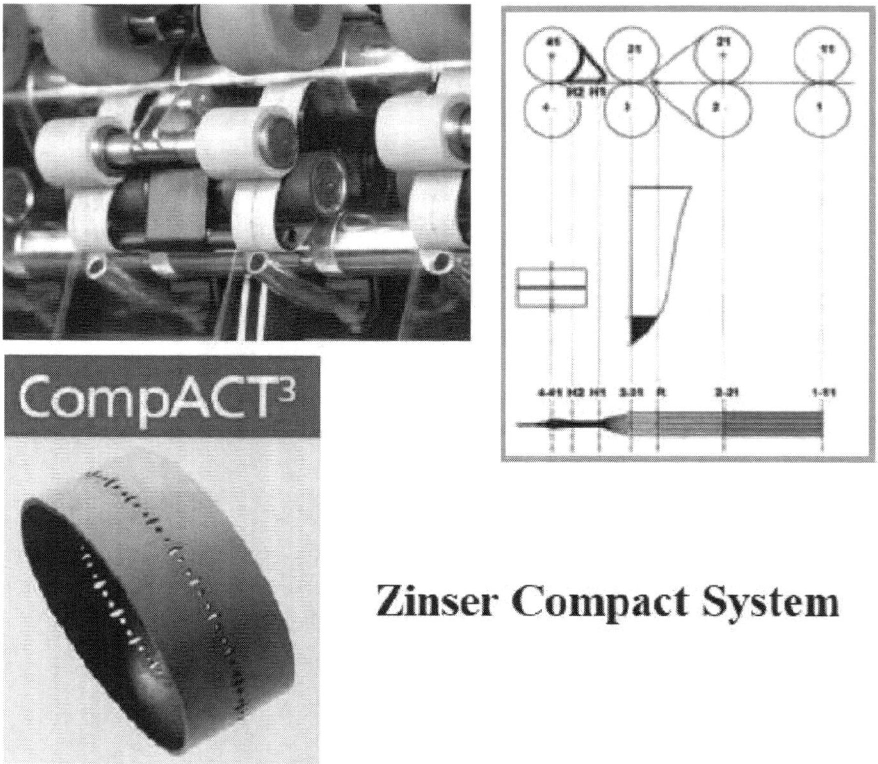

Figure 3.16 Zinser compacting system

3.6.5 RoCoS-magnetic Compacting System:

RoCoS stands for 'RotorCraft compact spinning' system which was developed by Hans Stahlecker of Rotorcraft Maschinenfabrik Switzerland and is available on ring frames of Lakshmi Machine Works (LMW). The RoCoS works without use of air suction. Magnetic mechanical compacting principle is used in RoCoS system. The RoCoS device consists of front bottom roller which supports the front top roller and delivery roller, in between these two rollers the ceramic compactor is placed. The supra magnets are equipped with ceramic compactor, which is pressed against front bottom drafting roller by supra magnet without clearance. It forms together with the bottom roller an overall enclosed compression chamber whose bottom contour, the generated surface of front bottom roller, moves synchronously with the strand of fibres and transport this safely through the compactor. According to Stahlecker , RoCoS 1 is suitable for 100 % cotton, cotton blends and 100% synthetic fibres with maximum staple length of 60 mm. RoCoS 2 is suitable for 100% wool,100 synthetic and wool/synthetic blends having minimum staple length of 50 mm. As RoCoS does not requiring air suction, air piping, perforated drum or apron and therefore there is no extra power and maintenance is necessary.

The RoCoS device (Please refer figure) consists of a cylinder (1), front roller (2), delivery roller (3), the precision ground and with supra-magnets equipped ceramic compactors (4), the supporting bridge (5), the yarn guides (6), and the top roller holders with the weighting springs (8).

The bottom roller has very precise flutes and radius exactly corresponding to the compactor radius. The bottom roller (1) supports the front roller (2) and the delivery roller (3). The precise magnetic compactor (4) is pressed against the cylinder. A and B are the two nips between which the compacting takes place. The magnetic compactor (4) as shown in Figure is pressed against the cylinder without any clearance against cylinder (1), thereby forming with the bottom roller an enclosed compression chamber where the bottom contour, i.e. the generated surface of the cylinder (1) moves synchronously with the strand of fibres and transports these fibres safely through the compactor. Therefore in the chamber formed, the compacting of the fibre bundle takes place, due to magnetic forces. The condensation of the fibres takes place to such a degree so that the formation of the spinning triangle is prevented while twisting of the fibres takes place.

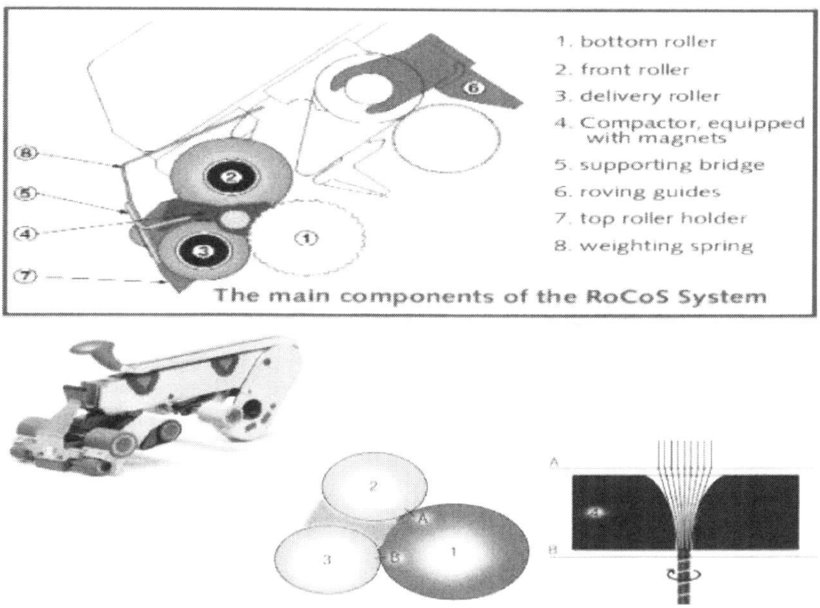

Figure 3.17 RoCOS Magnetic compacting system

Advantages of Compact Spinning:

Improvement in tensile properties, reduced hairiness as well as improved regularity of a yarn are the key benefits of compact spinning. Here some of the advantages of compact spinning are discussed. Advantages in Spinning:

1. As maximum number of fibres are integrated into yarn body during spinning, so better utilization of fibres at the same time less fly generation and clean atmosphere is spinning department.

2. Yarn twist can be reduced by 10% while maintaining the same strength as the conventional ring yarn. Therefore, it is possible to increase the machine speed which ultimately results into increased production.
3. The weak point in the spinning zone (spinning triangle) is eliminated, the end breakage rate is considerably reduced which again leading to higher machine efficiency.
4. 4. Increased strength and breaking elongation of yarn due to less protruding fibres and improved orientation of fibres, which leads to full realization of fibre strength.
5. Appreciable reduction in hairiness due to virtually elimination of spinning triangle.
6. Less expensive raw material can be used to produce good quality yarn.
7. Significant reduction in IPI, results better yarn quality.
8. Singeing can be completely eliminated.
9. Noil % at comber can be reduced as short fibres are better integrated into yarn body during spinning.
10. The improved characteristics of compact yarns give higher yarn sales price.

Advantages in Winding:

Due to end breaks in spinning, improved winding efficiency as few clearer cuts. Waxing of yarn can be eliminated.

Advantages in Twisting:

Lower twist can be employed in doubling, to improve the strength of yarn. The systems like COM4twin, Elitwist saves the cost of doubling. Advantages in Weaving Preparation & Weaving:

1. Better packing density of compact yarn gives better abrasion resistance and which leads to fewer end breaks in weaving. Also loom shed droppings and linting in knitting are reduced.
2. Degree of sizing can be reduced which reduces the sizing cost and the subsequent desizing cost.
3. Reduced end breaks in warping improves efficiency of warping.
4. Low end breaks in weaving improves weaving machine efficiency.
5. The compact spun yarn gives clearer cut contours in design.

Disadvantages of Compact Spinning:

1. Higher capital cost of the machine due additional condensing zone in drafting system.
2. Increased maintenance of condensing zone which adds to cost.

3.7 Developments and limitations of ring spinning

As indicated in the beginning of this module, ring spinning is the most versatile spinning process. It can spin yarns of a wide range of counts (from very fine to coarse) from different types of fibres (short as well as long staple fibres). The quality of ring spun yarns has been a benchmark against which the quality of yarns produced on other spinning systems is judged.

The basic principle of ring spinning has not changed much since its invention by Thorpe in 1828. But, there have been numerous developments of the ring spinning system, particularly in the short staple sector. These developments include:

- Automatic doffing of full cops (bobbins)
- Linkage to roving frame
- Linkage to automatic winding machine

Developments of worsted ring frames are relatively slow, because the market of worsted spinning is much smaller than short staple spinning. Therefore, the incentive for manufacturers of worsted ring frame is not very high. Over the years, the ring spinning system has also been modified to improve the properties of ring spun yarns. Examples of such modifications include:

- Sirospun (see the reading "Sirospun - A yarn with character" by Waldauser)
- Compact spinning (see the reading "The Suessen Elite Spinning system for long and short staple fibres", courtesy of Suessen, Germany).

Ring spinning also has several major limitations. These limitations include:

- High power consumption
- Small package size
- Low production rate

Staple spinning is basically about twist insertion. In ring spinning, twist insertion requires the rotation of the whole yarn package on the spindle. About 95% of the power used in ring spinning is consumed by rotating the yarn package to insert twist. This leads to the relatively high power consumption for the ring spinning systems. The package size is limited in ring spinning due to the need to reduce the balloon height and yarn tension, as discussed in the section on the physics of ring spinning. Perhaps the most serious limitation of ring spinning is its low production rate. We already know the relationship between yarn twist level, spindle speed, and yarn production or delivery speed. According to its end-use, a ring spun yarn needs to have a certain level of twist, which is determined before spinning is started. From the production equation, we know that the only way of speeding up the yarn delivery speed or production rate is to increase the spindle rotation speed. We already know that any increase in spindle speed will lead to significant increase in yarn tension, hence the possibility of endsdown.

In addition, with the increase in spindle speed, the traveller speed increases. This increases the friction between the traveller and the ring. Considerable heat is generated because of this friction, which may result in traveller burning during spinning. Because of these, spindle speed cannot be increased at will, and yarn production is limited as a consequence. Currently, the maximum spindle speed for short staple ring spinning is about 25,000 rpm, and that for long staple ring spinning is about 15,000 rpm.

Chapter 4

The Rotor Spinning Process

4.1 Introduction

In the production of short staple spun yarn, the rotor spinning technique is an excellent modern alternate to the old classical ring spinning method due to its high productivity advantage over the ring spinning. The rotor spinning is not only highly productive but also very cost effective at the same time. In addition to short staple spinning, long staple rotor spinning machines have also been manufactured only to produce coarse yarns from long staple manmade fibres.

However, wool fibres because of their scaly surface, crimp and natural greases are not possible to spin using this technique. Rotor spinning belongs to the family of open-end (OE) spinning. Open-end spinning systems are designed to overcome some of the problems associated with ring spinning. As discussed in the previous topic, twist insertion in ring spinning requires the rotation of the whole yarn package. In open-end spinning, only an end of the yarn is rotated to insert twist, which consumes much less energy than rotating a yarn package. The most successful examples of the open-end spinning concept are the rotor spinning and friction spinning systems. This topic discusses rotor spinning. Friction spinning will be discussed in the next topic. It is only logical, then, that open-end spinning will be gauged by its performance in direct competition with the older, proven, and accepted system.

Rotor spinning was commercially introduced in 1967. In its early years, the rotor spinning was mainly used to produce coarse low quality yarns. However, with modern developments, it is now possible to produce high quality yarns up to counts of 40 Ne. Furthermore, the rotor speed has been successfully increased from early range of 40,000 rpm to 150,000 rpm. These latest developments have given rotor spinning technique, the capability of producing high quality yarns at a productivity rates of up to 10 times higher than and as low as one third of the cost as compared to that of ring spinning technique. The biggest limitation faced by rotor spinning is its inability to produce very fine yarns that otherwise can easily be spun by using ring spinning.

Despite of this limitation, rotor spinning is a very good competitor of ring spinning and is capturing the market considerably in coarse yarn counts.

The rotor spinning technique is quite different from the conventional ring spinning. The differences are in the following aspects:

In ring spinning, the opened and cleaned fibre mass is converted into a strand form and a yarn is created by successive drafting and twisting at various spinning stages without disturbing the linear strand form of the fibres. Whereas in rotor spinning, the strand of fibres is first separated by vigorous drafting and is then recollected and twisted in a component of the machine called as a rotor.

The feed to the rotor frame is given rarely in the form of card sliver or most commonly in the form of drawn sliver. Since rotor spinning is a true sliver to yarn conversion as compared to ring spinning where no additional roving process is required, the productivity rates are much higher; it covers less floor space and is less labour extensive. In ring spinning systems the package formation i.e. winding and twisting is an integral part of the spinning system whereas in rotor spinning, the package winding is quite separate from drafting and twist insertion.

The rotor spinning produces fewer spinning faults like thick and thin places as compared to ring spinning, so rather than carrying additional winding and clearing process after spinning, the winding section of rotor frame directly produces suitable large packages required for subsequent processing with built-in clearers to remove spinning faults. This further makes the rotor spinning more economical and more productive. Instead of using the classical roller drafting, the rotor frame utilizes dispersion drafting technique. The method of twist insertion is also different. In rotor spinning twist is inserted by the rotation of the rotor whereas in ring spinning the rotation of the traveller around the ring inserts twist.

4.2 History of Open End Spinning

The global demand for spun fibre is huge. Converting raw fibre to yarn is a complicated process. Many manufacturers compete to provide the spinning machines that are essential to meeting the demand by delivering increases in spinning productivity and additional improvements in yarn quality. Over the past three centuries spinning technology has been continuously improved through thousands of minor innovations, and occasional major advances that have collectively increased the quality and lowered the cost of producing yarn dramatically. The industrial revolution which began in the eighteenth century produced many important innovations in the textile industry. Those related to the spinning of fibres are listed below:-

1764: James Hargreaves invented the 'spinning jenny'.

1769: Richard Arkwright invented the water-powered spinning frame.

1779: Samuel Crompton combined the two systems in his 'spinning mule'.

1828: Thorpe, Jenk and Mason created the prototype 'ring frame'.

1937: Berthelsen developed a relatively perfect open end.

1965: Czech KS200 rotor spinning machine was introduced at 30000 rotor rpm.

1967: Improved BD200 with G5/1 Rieter were presented with first mill of OE coming under production

1971–1975: There was a considerable increase in machine manufacturer and newer and improved version of machines were launched with increased speed at 100000 rpm.

1975: Also witnessed first automated machine from Suessen equipped with Spincat and Cleancat which opened up the industrial rotor spinning breakthrough.

1977: Witnessed Schlafhorst with Autocoro machines which made a mark in open end market.

Historical perspectives of rotor spinning

Compared with ring spinning, rotor spinning is a relatively new spinning technology that has not yet reached its maturity. A brief chronology of rotor spinning developments is listed below:

1937 The first idea and basic rotor patented by Berthelsen (Denmark).

1951 Meimberg (Germany) developed the invention further and built the first spinning models.

1965 Rohlena and his group (Czechoslovakia) found the correct combination of spinning elements and showed the first commercially functional units in Brunn.

1967 The Czech firm ELITEX exhibited its rotor spinning machine (BD200) near the international textile machinery exhibition (ITMA) in Basel, Switzerland. The machine had a rotor diameter of 75 mm, a rotating speed of 25,000 rpm, and a high twist multiplier (TM=6, or a twist factor of about 5740 tpm. tex).

1970 First sale of BD200 in the West

1971 Invention of the twin disk rotor drive allowing higher rotor speed (Suessen). With smaller rotors (60 mm), the rotor speed increased to 35,000 rpm.

1978 Introduction of 40-50 mm diameter rotors, improved spin box geometries, lower yarn twist possible, first automatic yarn piecer and package doffer fitted on the rotor spinner.

1989 Smaller rotors with speeds of 100,000 rpm.

1992 Quality yarns as fine as 13-15 Tex produced commercially at rotor speeds up to 120,000 rpm.

1999 Rotor speeds up to 150,000 rpm possible.

The development of rotor spinning technology continues today. The ultimate aim is to produce rotor spun yarns that match the quality of comparable ring spun yarns, but at a fraction of the cost of ring spinning.

4.3 General concept of open-end spinning

Open-end spinning is a relatively new concept of spinning. The basic principle of open-end spinning is illustrated in figure 2.1. Like ring spinning, open-end spinning involves the three basic steps of drafting, twisting and winding-on.

Drafting:

Very high draft is used to attenuate the feed sliver (not roving) into individual fibres. Such a high draft is usually by means of pinned drafting (with toothed rollers) rather than by roller drafting. Because of the direct sliver feed, there is no need to convert the sliver into roving first before spinning, which is necessary in conventional ring spinning.

Twisting

The individual fibres are collected at the yarn open-end and twist is then inserted at the yarn open-end. Since only the yarn open-end is rotated to insert twist, open-end spinning is much more energy efficient than ring spinning, which requires the rotation of a massive yarn package to insert twist. In addition, the twist insertion rate in open-end spinning can be very fast. For a given yarn twist level, this translates into fast yarn delivery speed or high production rate.

Winding-on

In open-end spinning, twisting and winding are separate operations so that yarns can be wound onto a large yarn package. In ring spinning, the package size is restricted and the yarns from the many small packages need to be joined up to make up a large package. In summary, openend spinning has the following major advantages compared to ring spinning:

- elimination of roving stage
- high productivity and low energy consumption
- large package size Now that we know the basic principle of open-end spinning and its advantages, we can proceed to discuss the details of rotor spinning. As mentioned in the introduction, rotor spinning is a successful example of the open-end spinning concept. We start with a brief account of the history of rotor spinning.

4.4 Principle of Rotor Spinning

The principle of open-end spinning (alternatively known as 'break spinning' or 'free fibre spinning'.) is illustrated in Figure 4.1. The essential features of the process may be summarized as: opening, transport, alignment, overlapping, and twist insertion.

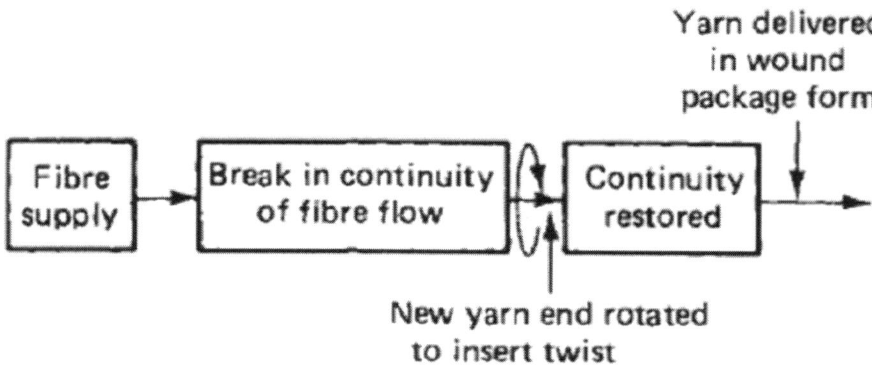

Figure 4.1 Schematic representation of the principle of Rotor spinning

The essential feature of the rotor spinning system is the separation of the functional stages of fibre sliver opening and yarn formation, respectively imparting twist and winding up the yarn.

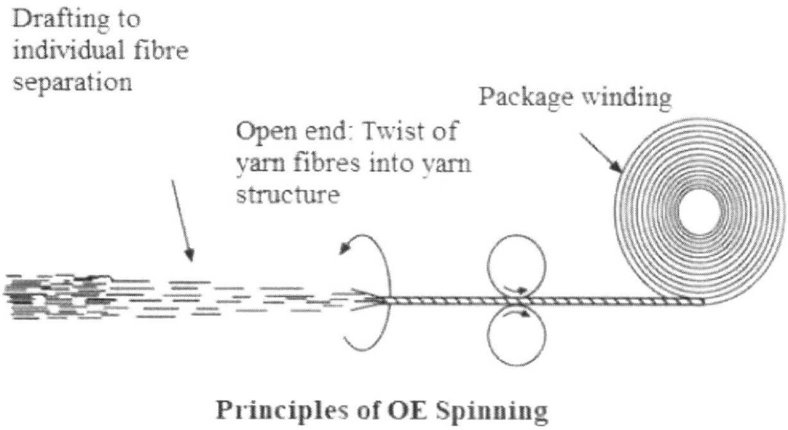

Figure 4.2 Schematic illustration of the principle of Rotor spinning

In order to achieve this the fibre bundle has to be interrupted at one point at least. This occurs between the functional stages of opening the draw frame or card sliver into individual fibres and subsequently combining these fibres in the collecting groove of the spinning rotor, the twisting device of the rotor spinning system.

Since the individual fibres are released from a compact fibre bundle during transport between the opening roller and the rotor collecting groove and are only combined again in the rotor groove, we can here refer to an open yarn end.

4.5 Basic methods of open-end spinning

Although many individual OE spinning devices have been invented, they may be classified into the following groups:

1. Vortex assembly
2. Axial assembly
3. Discontinuous assembly
4. Friction spinning
5. Rotor spinning

Rotor spinning is an open end process which generates a genuine yarn twist. In this case the component imparting the twist is the rotor, which twists the thread around its axis. The resulting yarn twist is the decisive factor for yarn tenacity.

However, in order to maintain the spinning process, i.e. integrate the fibres in the rotor groove, a spinning twist is required, which as a rule must be higher than the yarn twist required for yarn tenacity.

This means that an additional twist must be imparted to the radial section of yarn (imparting false twist). This false twist is imparted by the unrolling motion of the yarn on the draw-off nozzle, which is therefore much more than a thread guide. Depending on spinning conditions, the false twist can be up to 60 % of the set yarn twist.

Functions of Rotor Spinning Process

Following are the important functions of the rotor spinning process:

Opening & Attenuation, the fibres in a sliver form are vigorously opened up into individual fibre form using opening roller having its surface covered with sharp teeth or spikes. The vigorous opening of the fibres in this way also helps to reduce the linear density of the material. The amount of reduction in the linear density depends upon the yarn count to manufacture and can be controlled by the rate of feeding and degree of opening carried out. The drafting technique used here also called as dispersion drafting is quite different from the roller drafting technique used in ring spinning.

Cleaning, during the opening of the fibres into individual fibre form, cleaning is carried out by removing trash particle and dust particles using trash removal devices. The cleaning at the rotor spinning is optional and is only used for more dirty cotton. It is very important that fibres reaching the rotor must be free from any trash particles or fragments because they can rapidly clog the rotor causing thick and thin places, neps and uneven compactness in the yarn.

Improving Evenness, back doubling is used to make an even and homogenised rotor spun yarn. The term 'Back Doubling' refers to opening of the fibres in individual form and recollecting them at the rotor for yarn formation. The evenness achieved by this method is far more than that achieved at the ring spinning.

Twisting, in order to impart strength to the yarn twist is vital. Twisting is carried out by collecting the individually opened fibres and subjecting them to high speed revolving rotor. The centrifugal forces caused because of high rotary speed of the rotor cause the fibres to get collected around the wall of the rotor in form of a yarn and with one complete rotation of the rotor, one twist is imparted.

Winding, the yarn produced by opening (drafting) and twisting is finally wound on finished packages. Unlike ring spinning where the package has lot of defects and also the size of the package is very limited due to ring / traveller arrangement, the winding portion of the rotor frame produces suitable bigger packages that can be directly used for next stage of processing i.e. weaving or knitting. In modern rotor machines, electronic clearing devices and automatic yarn piecing arrangements have also been provided.

Working Areas of the Rotor Frame

The rotor spinning process can be divided into following areas:

1. Feeding
2. Opening & Drafting
3. Fibre Transport
4. Fibre Reassembly
5. Twisting

6. Winding

Feeding Zone

The feed to the rotor machine is either done in the form of card sliver or drawn sliver. Most commonly drawn sliver of either first or second passage is used as a feeding material to the rotor frame. The sliver is fed to the opening zone with the help of a feed roller. The rotation of the feed roller grips the fibres and presses is against the presser plate so that a controlled feed may be given to the opening zone of the machine.

Opening & Drafting Zone

The separation of fibres from sliver to individual form is critical for uniform yarn formation. If fibres are not delivered to the rotor with proper individualization, the resultant rotor yarn will lack orientation and uniformity. The opening of the fibres is carried out with the help of a rotating opening or combing roller whose surface is covered with saw teeth. As the sliver is fed in the opening region, the fibres are caught by the teeth of the opening roller. Here the centrifugal forces and the aerodynamics of the system, transports the fibres from the teeth of the opening roller to the fibre transport tube. In the fibre tube an air stream is provided that further does the opening and ultimately deposits the fibres in the rotor.

As the fibres are separated from one another, the opening roller also performs cleaning. Cleanest possible fibres should be fed to the rotor machine, however at such high degree of opening, some fine trash particles and dust will be generated. The trash extraction unit is designed so that lighter fibres are allowed to be carried away with the air stream into the fibre transport tube while the heavier trash particles will directly fall due to their weight into the trash extraction duct. An illustration showing a typical rotor spinning process is shown below:

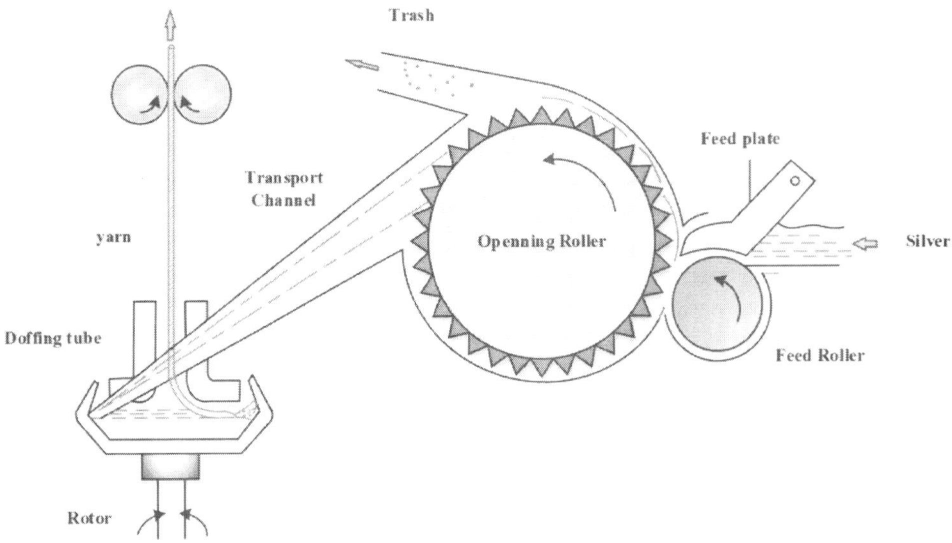

Figure 4.3 Schematic illustration of the rotor spinning process

This extensive opening of the fibres results in drafting of the input sliver material and reduces its linear density. The type of drafting employed here is called as dispersion drafting which is quite different from the roller drafting technique used at the ring spinning frame. Usually a draft of range of 100 to 200 is applied during opening.

Fibre Transport

The fibres opened up by the opening roller must be transferred to the rotor without getting disoriented. This is achieved by using a specially designed transport tube. The fibres are transported by this tube with help of air currents. The transport tube is slightly tapered to accelerate the air currents and fibres so that hooked surfaces caused by teeth of opening roller may be straightened out and fibres can be oriented.

Fibre Reassembly

The fibres coming from the transport tube are accumulated inside the rotor. Rotor is the main component of the rotor spinning process. Rotor is just like a small metal cup with inclined walls having a conical shape. The inclination is essential so that the fibres coming from the fibre feed tube can slide downwards. The inner surface of the rotor is called as collecting groove. The diameter of the collecting groove also called as the rotor diameter depends upon the speed of the machine and fibre length. As a rule, the rotor diameter should never be less than 1.2 times the staple length of the fibres being processed.

The rotor is generally made of aluminium or steel having its surface covered with diamond particles covered in Nickel to reduce wear and increase its working life. A shaft is fixed to the rotor so that when it is rotated, the rotor will also rotate. The rotor rotates inside a stationary cover or housing with the help of a bearing. In the middle of the rotor a withdrawal tube or doffing tube with specially designed mouth piece or navel is provided for the yarn to leave the rotor. Please refer figure 4.4

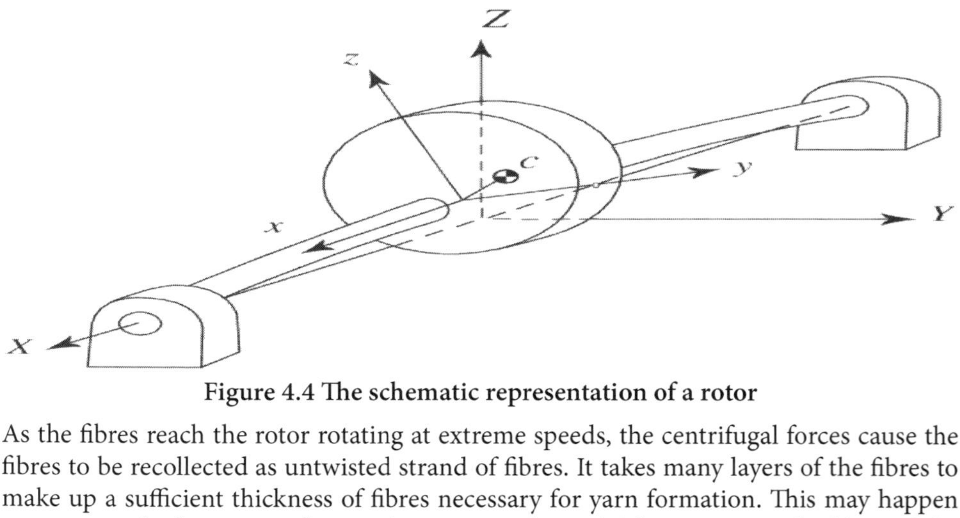

Figure 4.4 The schematic representation of a rotor

As the fibres reach the rotor rotating at extreme speeds, the centrifugal forces cause the fibres to be recollected as untwisted strand of fibres. It takes many layers of the fibres to make up a sufficient thickness of fibres necessary for yarn formation. This may happen over a period of many revolutions (approximately 100). The numerous doubling of the layers of fibres provide blending called as back doubling. This back doubling makes the rotor spun yarn very even with fewer thick and thin faults as compared to ring spun yarn.

Twisting

The twisting occurs with the action of rotor, navel and the take-up rollers.

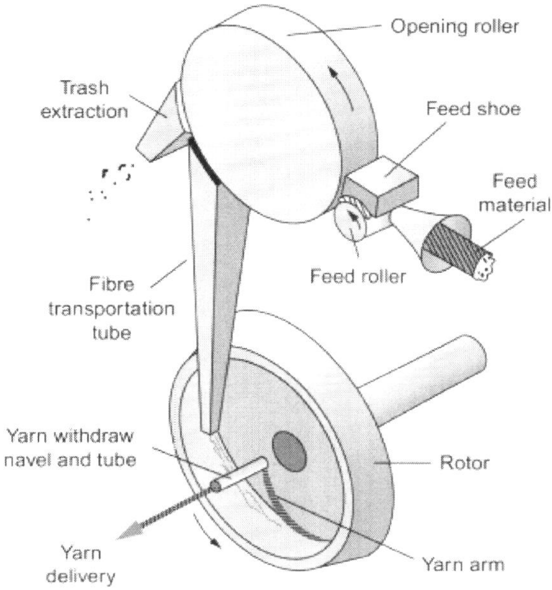

Figure 4.5 Schematic representation of the rotor

Individual fibres emerge simultaneously from the fibre feed tube, they slide along the inner wall of the rotor and are collected around the collecting groove. In this way a continual fibre ring is built up in the groove and this process is called as the back doubling. As the sufficient number of fibres has reassembled inside the rotor wall, further fibre feed will choke the rotor.

However rotor's special aero-dynamical design and due to excessive centrifugal forces acting on the fibres, the yarn is allowed to extend from the rotational axis of the rotor to its outer surface and ultimately carries the yarn to the navel from where it can be drawn forward with the help of a take-up rollers.

Each revolution of the rotor inserts one twist in the yarn. The twist travels from the back of the navel to the point where the fibres leave the rotor. When the twist in the yarn reaches the maximum level, the yarn end begins to rotate on its axis. It is necessary at this point to carry the yarn perpendicularly out of the navel with the help of the take-up rollers. Mathematically the amount of twist is calculated by the following formula:

Turns per inch = Rotor Speed (rpm) / Delivery speed of yarn ("/min)

However the calculated twist always comes out to be 15 to 40% lower than the actual twist imparted on the yarn. This is due to the fact the fibre slippage occurs during the actual twist insertion.

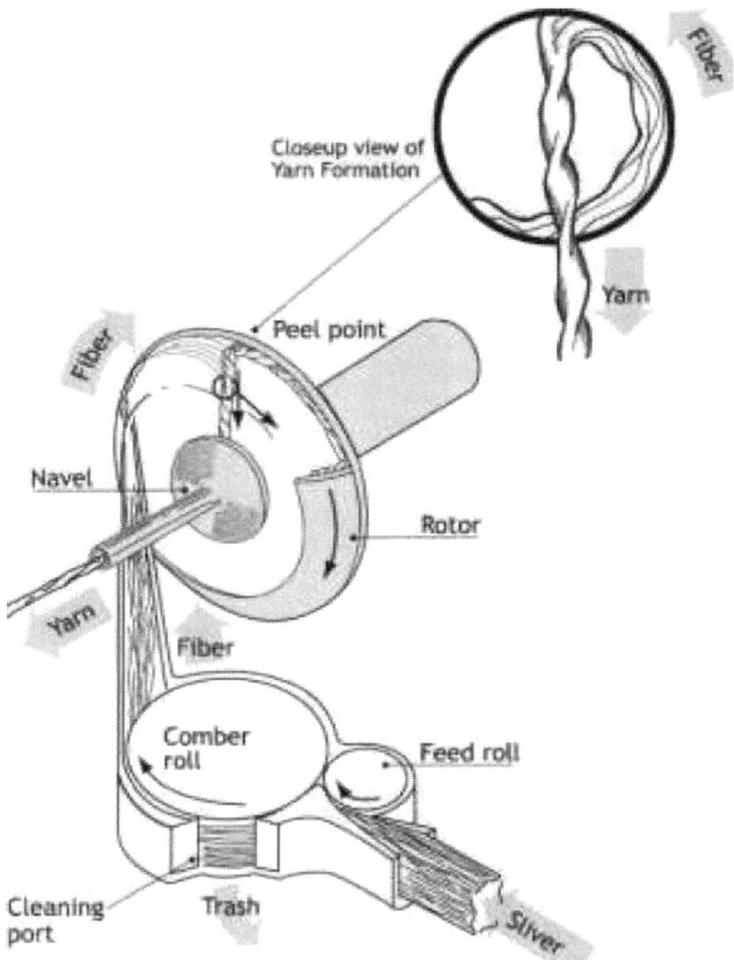

Figure 4.6 Rotor Spin box

Please refer to the figure 4.6 a for the configuration of the rotor spin box elements of a rotor spinning machine. Since the yarn travels through the navel and the doffing tube with a rolling action touching their surfaces, false twist is inserted in addition to the real twist inserted at the rotor. So the total twist inserted in the yarn is the sum of the real twist and false twist. The false twist provides more stability to the yarn however during the package winding the false twist is automatically removed and the final yarn only has the real twist in it. The false twist does have a considerable impact on the yarn characteristics. At higher rotor speeds due to more centrifugal forces, the amount of false twist inserted during yarn formation will be more. The increase in the false twist increases the amount of wrapper fibres in the yarn causing the yarn to have more protruding fibres and a harsher look and feel.

The heart of the open-end process is a rotor (see Figure 4.7), wherein fibres can be collected and then drawn off as a yarn. For short staple spinning, most rotors are 31 to 56 millimeters in diameter and may contain a shallow "U" or "V" shaped fibre alignment groove around their periphery. In open-end spinning, the rotor rotation provides the twisting force.

Figure 4.7 Rotors

Twist has traditionally been inserted into yarn by rotating the package upon which the yarn is being wound. In the case of open-end spinning, the twisting force is generated by the rotation of a rotor and is transmitted by friction to the fibres that make up the tail of the newly-formed yarn. As this twisting tail comes into contact with other fibres, it collects them. Once this process is started, it is self-sustaining, and yarn can then be drawn out of the rotor continuously.

In order to prevent twist from being transmitted throughout the length of the fibres that are available for collection into yarn, it is necessary that these fibres not be in any significant frictional contact with one another. It is from this requirement, that the supply fibres not be in intimate frictional contact, that open-end spinning derives its descriptive name. This lack of contact allows true twist to be inserted into the yarn, and at the same time, prevents twist from being transmitted throughout the fibre supply, which would result in instant stripping of the rotor.

Since the yarn passes out of the navel in a rolling action touching its surface, the surface and design of navel has a great influence on levels of false twist and ultimate quality of the yarn. Based upon the type of the yarn to produce, different shapes and roughness of navels are available as shown below: Please refer figure 4.8.

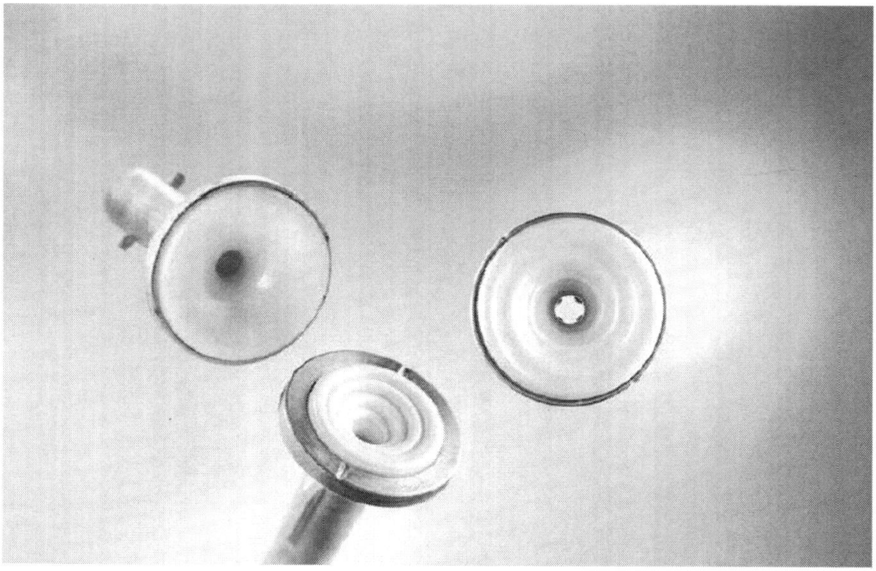

Figure 4.8 Navels (Take off nozzles)

Differences: The basic difference between ring-spun yarns and open-end spun yarns is in the way they are formed. The former produces yarn by inserting twist into a continuous ribbon-like strand of cohesive fibres delivered by the front rolls, while the latter forms yarn from individual fibres directly by collecting them from the inside surface of a rotor by twist forces. Thus, for comparison, it could be said that a ring yarn is formed from the outside in, while open-end yarn is formed from the inside out. This is shown in Figure 4.9.

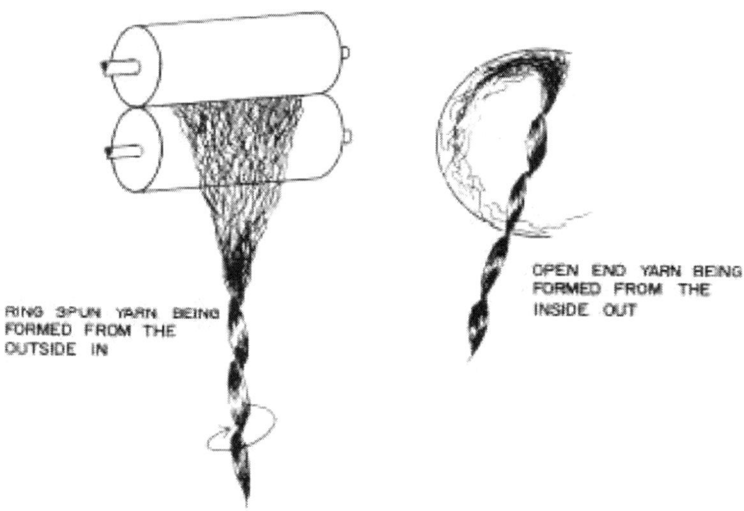

Figure 4.9 Comparison of ring and rotor spinning yarn formation

The elements basic to production of open-end yarns are somewhat different from ring spinning (Figure 4.8). They are:

a) fibre supply,

b) drafting system,

c) fibre collection and alignment,

d) twist insertion -- yarn formation, and

e) package winding.

The fibre supply used by all open-end machinery is in the form of sliver, either directly from the card or from the draw frame. This being the case, it is evident that the roving process long associated with ring spinning is no longer needed.

When this fact is considered, it becomes apparent that the total draft for a given yarn produced on an open-end machine will be greater than that of a ring frame producing the same yarn. For instance, the draft of an open-end frame producing 24/1 from 60 grain sliver would be 172.8, while the draft on a ring frame producing the same 24/1 yarn from 0.80 hank roving would be 30. Thus, the overall draft of an open-end frame is likely to be quite high; as a result, open-end equipment manufacturers decided to abandon roller drafting as being too cumbersome to use to develop the high drafts required for rotor spinning.

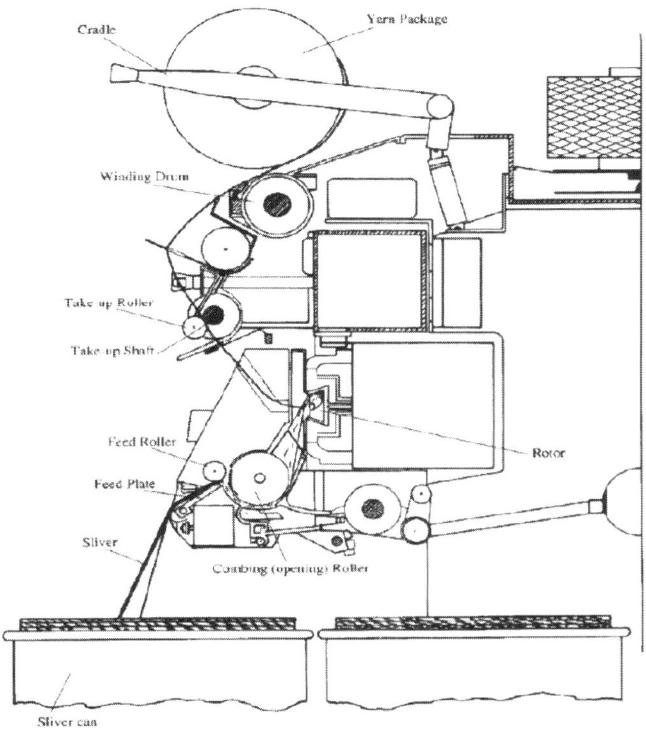

Figure 4.10 Cross sectional view of a rotor spinning machine

In place of roller drafting, a simple wire or pin-covered cylinder several inches in diameter and about an inch wide is used (Figure 4.11). This beater, or combing roll, as it is normally referred to, works just like a card lickerin; it is, in fact, covered with metallic card wire, or possibly steel pins. The rotational speed of a combing roll ranges from about 3,000 to 10,000 rpm with 4,000 to 8,000 being most common.

In actual operation, the supply sliver is presented to the rotating teeth of the combing roll by action of a feed roll/feed plate mechanism. When the position is started, a brake is released that allows the feed roll to pull sliver between itself and the feed plate, and thus present a beard of fibres to the combing roll. (Figure 4.10.)

Figure 4.11 Opening rollers

There is a very high draft between the feed roll/feed plate and the rotor. This draft normally will vary in a range from 1,000 to 40,000 or higher, depending on the count produced, draft, and required twist. This high draft delivers individual fibres and/or small individual fibre groups to the rotor where they are deposited randomly around the inside of the rotor over a period of many revolutions. This deposited fibre mass has very little fibre-to-fibre cohesiveness, and it is this fact that makes open-end spinning possible.

The combing roll combs out fibres and carries them to the fibre transport duct that connects the combing roll chamber to the rotor chamber. (Figure 4.12.) A combination of centrifugal force and air suction from the rotor chamber tends to strip the fibres away from the combing roll and send them into the fibre transport duct.

The fibre transport duct is aimed tangentially at the rotor's fibre alignment groove. This duct exit usually is located very close to the groove to prevent newly-arrived fibres from being collected by the newly-formed yarn before ever reaching the rotor alignment groove. Some manufacturers choose to align the fibre delivery tube so that fibres entering the rotor first contact the rotor wall just above the fibre groove, and then slide down the wall into the groove.

These manufacturers believe that better fibre alignment is obtained using this method. Other machinery manufacturers do not use the fibre delivery duct to direct the fibres into the rotor groove, but rather use a separator plate to direct newly-arrived fibres into the fibre alignment groove.

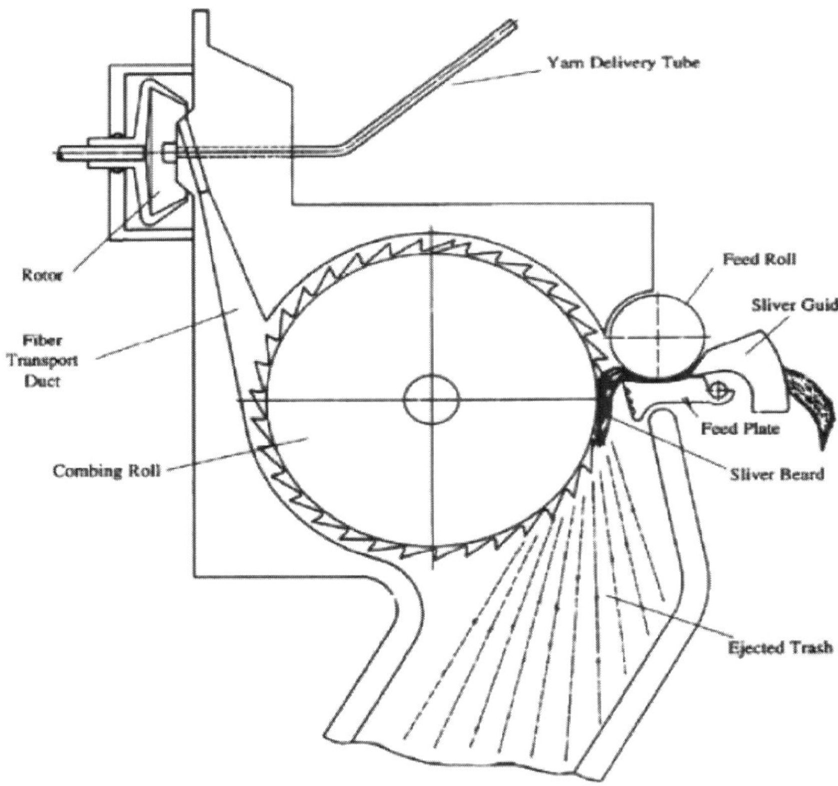

Figure 4.12 Cross sectional view of the rotor box

Some machinery builders have designed a trash extraction feature into their combing roll drafting systems. This feature is illustrated in Figure 4.13. The system is designed so as to allow the lighter fibres to be carried by air currents and the combing roll teeth safely across the port while the heavy trash particles, by reason of their mass, will deflect through the opening and out of the system.

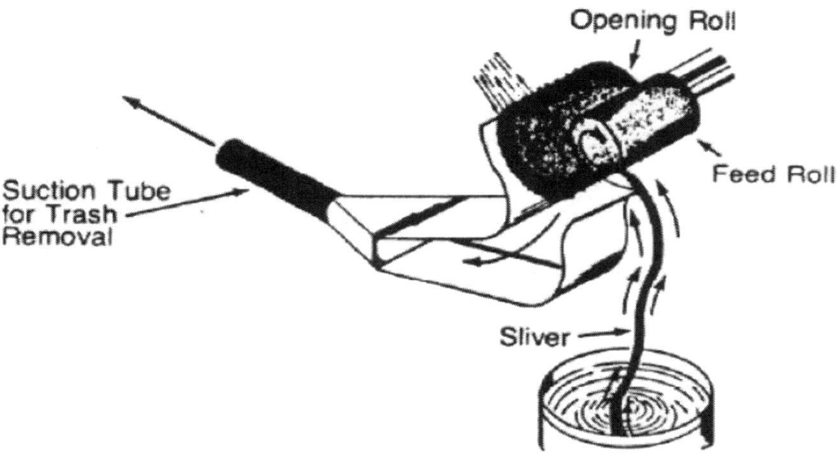

Figure 4.13 Dust extraction arrangement

Open-End Yarn Formation: In rotor spinning, the yarn is formed inside the rotating rotor from a continuous stream of individual fibres arriving from the combing roll. This spinning action can be explained as follows. Please refer to Figure 4.14, which is a schematic diagram of the inside of a rotor.

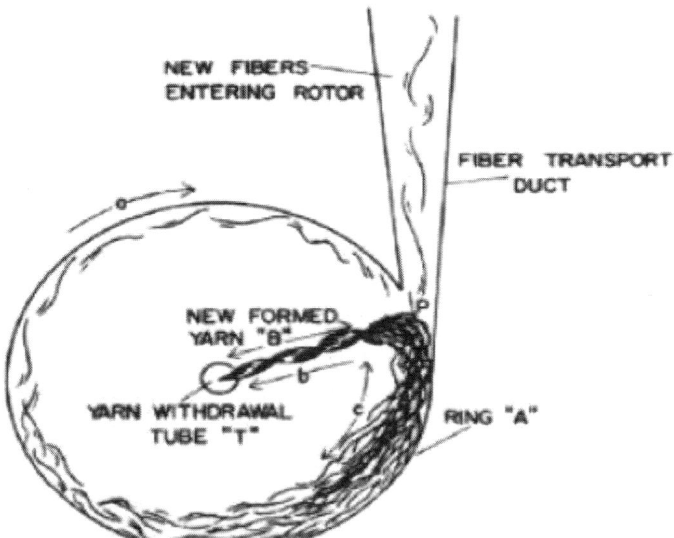

Figure 4.14 Schematic illustration of yarn formation inside the rotor

Ring "A" represents the inside groove of the rotor where the fibres are collected. Ring "A" rotates in the direction of Arrow "a" at a fixed rate. Newly formed yarn is seen at "B" and moves in the direction of Arrow "b" to the yarn withdrawal tube "T" where it is drawn out of the rotor and wound onto a cheese. The new yarn is actually formed in Area "C" by twist collection of individual fibres. This area is known as the fibre binding zone. The point where the fibres leave the rotor surface is called the "peeling point" and is identified as Point "P." The peeling point advances in the same direction as the rotation of the rotor, shown by Arrow "a."

The rate of advance of this point is determined by the turns per inch of twist being inserted into the yarn. One turn of twist is theoretically inserted into the yarn each time the rotor makes one complete revolution. This being the case, it becomes obvious that rotor speed and production are directly related.

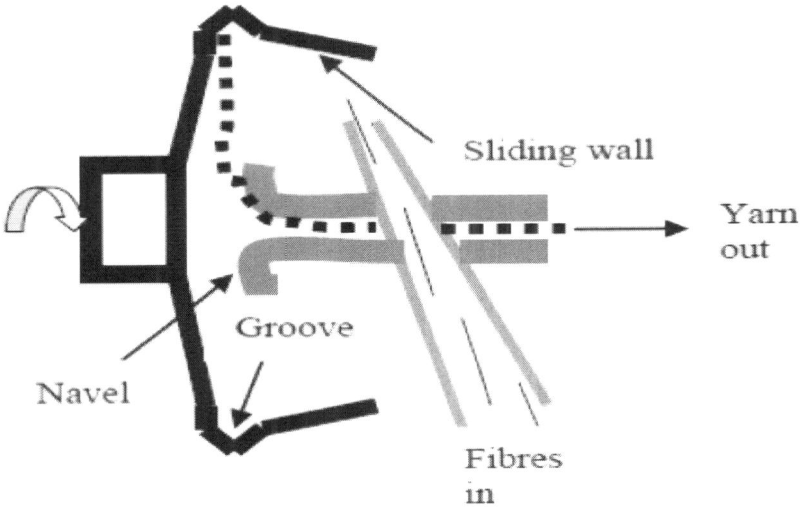

Figure 4.15 Schematic illustration of the inside of a rotor

Once a sufficient number of fibres is collected in the rotor groove, the fibres need to be taken out continuously otherwise they will soon clog the rotor groove. To do this, a 'seed yarn' is first introduced into the rotor through the navel (figure 4.15). Again, the centrifugal force throws the seed yarn into the rotor groove. As the rotor is rotating rapidly, the seed yarn rotates with it. This rotation traps the loose fibres at the end of the seed yarn. At this point, the seed yarn is pulled out, the fibres trapped to the yarn end are peeled off the rotor groove by the outgoing yarn. Since the peel-off point is rotating with the rotor, twist is inserted into the out-going fibres. In rotor spinning, the drafting mechanism consists of three main operations: (i) mechanical opening using an opening roll, (ii) air drafting using an air stream and transporting duct, and (iii) doubling mechanism. The mechanical drafting is achieved using a toothed opening roll. In order to minimize fibre disorientation, the airflow in the duct should have a velocity exceeding that of the surface speed of the opening roll. To obtain such an airflow, the inside of the rotor is run at a vacuum which may be achieved by designing the rotor with radial holes to allow the rotor to generate its own vacuum (self-pumping effect).

Winding

The yarn formed at the rotor leaves the rotor spin box and is finally wound into a finished package. Unlike the ring spinning frame where the size of the package is limited because of the ring / traveller arrangement, bigger cross wound packages are produced either in form of cheeses or cones. During the process of winding spinning faults are also removed to produce an even fault free packages that require no additional processes and can be directly used for weaving or knitting. Please refer figure 4.16.

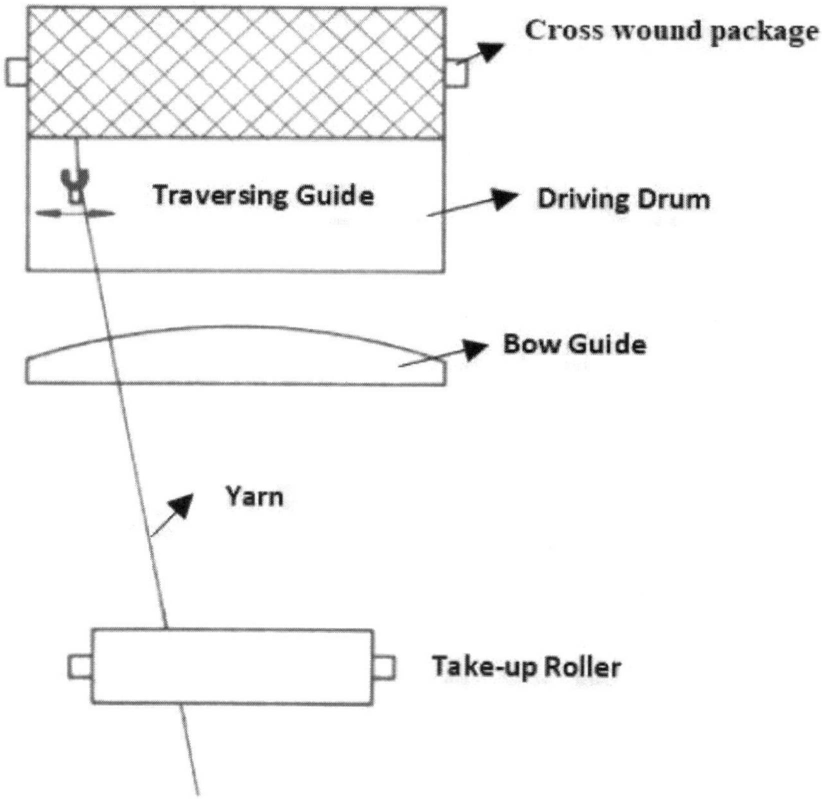

Figure 4.16 Schematic diagram of winding on a rotor spinning machine

The yarn produced at the rotor is continuously pulled by the pair of take-up rollers placed just after the doffing tube and ultimately winds the yarn on the package. The package is driven negatively by the frictional contact of the driving drum. To wind the yarn evenly throughout the package length, a linear traversing guide is used. However in case a grooved driving drum is used, this linear traversing guide is no longer used as traversing is performed by the grooved drum itself.

Winding is only possible if the surface speed of the yarn being wound on the package is more than that at which the yarn is delivered. This creates a tension called as the winding tension. The winding tension controls the compactness of the package. More the winding tension, more compact and hard packages are formed and vice versa. In case the package is to be dyed, the winding tension is kept low as compared where normal weaving or knitting packages are formed.

For knitting yarns, waxing devices are used to apply wax to the yarn so that the yarn is lubricated to reduce yarn to metal friction in the knitting needles so that the wear of knitting needles may be reduced.

Electronic Clearing & Automatic Piecing Devices

The modern rotor spinning machines are also equipped with electronic clearers to remove the spinning faults and automatic piecing devices to repair the broken ends of the thread. Since the number of thick and thin places produced by the process of rotor spinning is far less as compared to the ring spinning, most commonly one piecing device is used for 10 to 12 rotors.

The piecing devices either reciprocate in between the specified rotors or they move in an endless circular path around the rotors. As the piecing device finds a stopped rotor, in addition to automatic piecing, it will also open up the spin box and will clean the rotor with the help of a specially designed brush. The automatic piecer works on pneumatic principle similar to the one used in automatic winding machines. The piecing is carried using splicing technique with the help of air currents and air suction.

4.6 General Cotton Fibre Considerations for Successful Open-End Spinning

Cotton is a very unique natural fibre, possessing a wide variety of properties that make the fibre universally useful. Over the years, ring spinners have determined which cotton properties must be controlled to enable them to efficiently produce quality yarns. These properties are length, length uniformity, strength, micronaire and non-lint content.

These same properties are of paramount importance to open-end spinning also, but the order of importance varies considerably. Based on available data, the probable order of property importance for open-end spinning is as follows:

a) non-lint content,

b) strength,

c) micronaire,

d) length,

e) length uniformity.

The basic importance of each of these properties to open-end spinning follows:

Non-Lint Content: The non-lint content of the input sliver has a profound effect on both yarn quality and machine performance. Any non-lint particles that enter the rotor suspended in the fibre transport air are subject to centrifugal forces generated by the rotation of the rotor and sufficient to cause these particles to deposit on the inside surfaces of the rotor. Once deposited, there are only two ways these particles can be removed: (1) to be picked up by the yarn and carried out, and (2) to stop the rotor and clean it by hand.

The deposition of particles can be of either a uniform nature, which affects overall yarn quality and strength by filling up the fibre alignment groove, or it can be concentrated in one spot, caused perhaps by the jamming of a relatively large particle into the alignment groove. This type of loading problem will certainly result in a periodic defect in the yarn and very likely an end down. The nonlint content of input sliver to be processed on self-pumped open-end machines probably should not exceed 0.10% for best performance. For separately pumped open-end machines with combing roll cleaning ports, the non-lint content of the input sliver should be no greater than 0.25% for best performance. However, machines with built-in cleaning can tolerate levels in excess of 0.25% non-lint with some loss in machine efficiency and increased maintenance costs.

There are two reasons why the separately pumped machines can tolerate a higher non-lint content. The first is, of course, the incorporation of a cleaning port under the combing roll that functions like a fibre retriever under a card. The fibres are carried over the opening by virtue of their length and velocity while the heavy trash particles tend to deflect away from the fibre path and through the cleaning port.

The second reason is that since the rotor itself does not generate the air flow required to transport the fibres from the combing roll to the rotor, the path of the air flow is across the face of the rotor rather than through it. Thus, an alternate path is offered to fine dust particles and this results in an additional reduction, but not an elimination, of the non-lint deposition problem. The importance of controlling non-lint content cannot be overemphasized.

Micronaire: The weight or fineness of a cotton fibre is related to its surface area. The greater the surface area, the coarser the fibre. (Micronaire is defined as the average weight of one-inch lengths of fibre expressed in micrograms.)

To successfully spin yarn, a certain number of fibres are needed in the yarn cross-section. In other words, a certain minimum number of fibres must be delivered per unit length of a given yarn or else the ends down will be excessive. This number is not precisely fixed since it is influenced by factors such as fibre length, strength, and twist multiple. However, it is generally accepted that ring spinning requires a minimum of 60 to 80 fibres in the cross-section. 100 fibres per yarn cross-section was the accepted minimum on rotor spinning until recent years when advancements in machine design made possible numbers less than 100 Therefore, from an efficient operating point of view, it should be clearly understood that in finer counts, openend yarns are much closer to the critical red line requirements for minimum fibres in the crosssection than would be the case if the given yarn was being ring spun. This is graphically shown in Figure 4.17.

The reverse is true for very coarse counts. In other words, more fibres may be required than can pass by the combing roll and through the fibre transport duct to the rotor. For every openend spinning system, there is a maximum number of fibres that can pass through the fibre transport duct without choking the opening.

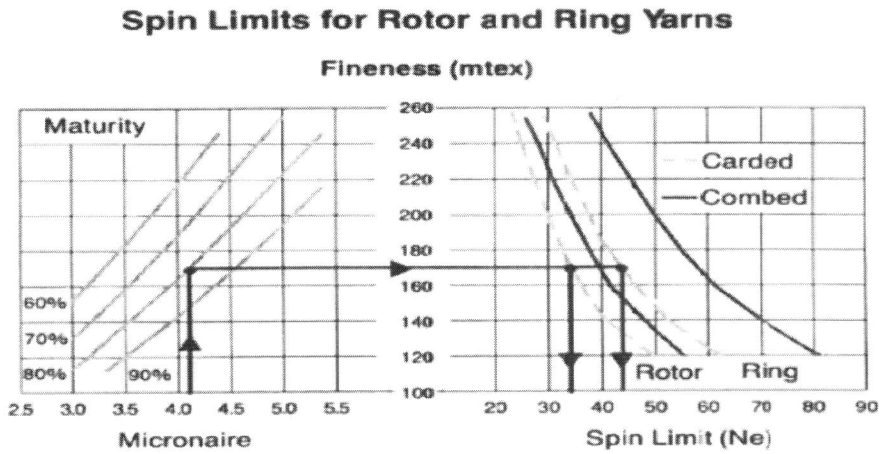

Figure 4.17 Yarn spinning limits for ring and rotor spinning

Generally speaking then, it should be appreciated that it is possible to spin finer counts or use a lower twist multiple by using finer mic cotton than is normally used in ring spinning. When using low mic cotton, neps can be controlled by using cotton with a high degree of maturity and by selecting the correct card wire.

Strength: There is a significant relationship between fibre strength and resultant yarn strength. Our tests have proven time and again that if a stronger open-end yarn is required, cottons with a higher strength should be utilized. This is graphically shown in Figure 4.18.

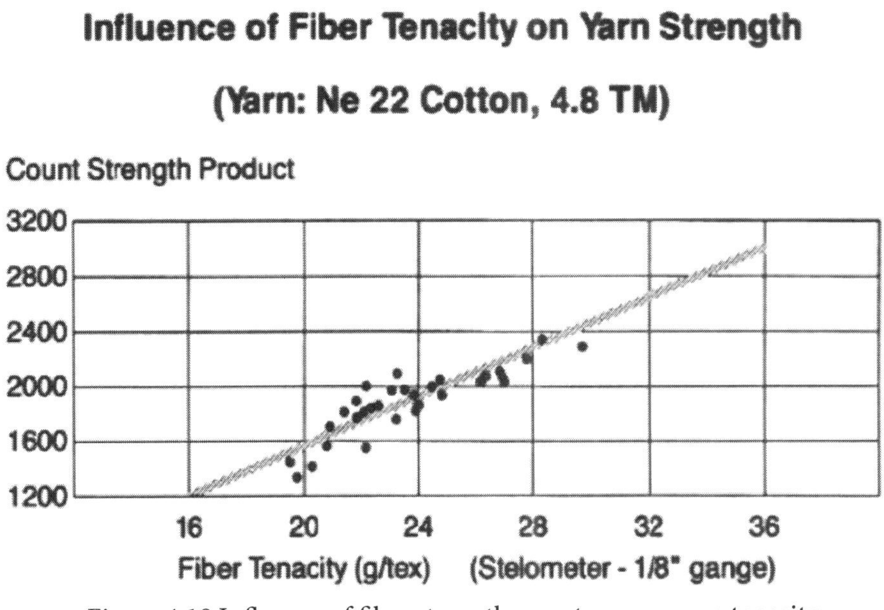

Figure 4.18 Influence of fibre strength on rotor spun yarn tenacity

Length: For any given rotor diameter, there is an optimum staple length for best strength, and although the use of longer staples will result in higher strength, it will do so at a rapidly diminishing rate. This is due mainly to the fact that as staple lengths increase, the opportunity for bridging fibres to occur increases. A bridging fibre by definition is one that spans the space on the inside surface of the rotor between the yarn formation or peeling point and the body of newly deposited fibres just behind it. The percentage of bridging fibres may be roughly calculated by dividing the staple length by the inside circumference of the rotor and multiplying by 100 to give percentage. Therefore, although staple length definitely has an effect on yarn strength, it tends to be greatly minimized by the open-end spinning process.

Open-end spinning is ideally suited for spinning of short fibres. A ring-frame drafting system has a minimum staple length requirement for proper fibre control to produce a quality yarn.

Once control is lost, the resultant yarn spins inefficiently, and the yarn appearance is poor. Since open-end yarn is formed by twist attraction of the rapidly rotating open-end, fibre control is no problem and, therefore, short staple fibre can be spun into more even and, in some cases, stronger yarn.

It is interesting to note that as staple length increases, open-end yarn becomes progressively weaker when compared to the same fibre spun into ring yarn. This differential may be as great as 25% to 30% on staples longer than 1-1/8". However, as the staple gets shorter, this differential decreases rapidly and, on some coarser yarns, the open-end yarn might even be slightly stronger. The above assumes comparisons of fibre properties on the same diameter rotor -- as new rotor configurations and sizes are studied, some change in fibre data/performance characteristics can be expected.

Even though fibre length ranks fourth on the rotor spinning list, it is still very important, especially when spinning fine counts. It can be observed from studies that when the staple length of a fibre is increased from 1" to 1-3/16", very little strength is added to the coarser yarn counts. On the other hand, the strength of the fine yarn counts increases considerably.

Length Uniformity: Length uniformity has always been considered of prime importance to ring spinning, and for best rotor spinning results, good length uniformity is also important, especially when spinning fine count yarns.

4.7 Structure of Rotor Yarn

The inner core of the rotor spun yarn is similar to a ring spun yarn. However the outer surface of the yarn has its unique structural build up owing to the method of yarn formation. During the yarn formation the fibres are reassembled with in the collecting groove of the rotor and are twisted to form a yarn. During the twisting of fibres into a yarn, portion of fibres instead of being twisted into the inner yarn structure are wrapped around the outside of the yarn. These fibres are called as wrapper fibres or bridging fibres as shown below:

The wrapper fibres make the rotor spun yarn harsher and hairy in appearance. They also contribute to increased needle wear in knitting and increased wear to loom components in weaving. Fewer the wrapper fibres present in the rotor yarn, more it will resemble to the ring yarn and the better the rotor yarn will be. The formation of wrapper fibres depends upon many factors, most important ones are:

Rotor Speed, with increasing rotor speeds, because of the centrifugal forces both the level of false twist and the yarn rotation at the navel increases and hence more wrapper fibres are wrapped around the main core yarn.

Rotor Diameter, with smaller rotor diameters, the wrapper fibres are wound fewer times around the yarn core making it more hairy but less bulky (more compact) as compared to rotor yarns spun on larger rotor diameters as shown in Figure 4.19 below:

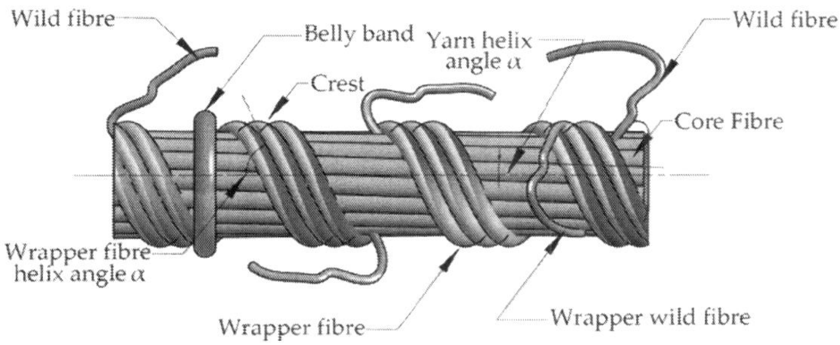

Figure 4.19 Rotor spun yarn structures

Fibre length, longer the fibres being processed, larger rotor diameters are required otherwise the fibres will get damaged. So, when larger rotor diameters are used at high speeds for longer fibres, the amount of wrapping fibres will increase.

Friction between fibre & rotor groove, more the fiction generated between the two, more wrapping fibres will be generated.

Surface of navel, rougher the surface of navel, more wrapping fibres are generated and vice versa.

The increase in the rotor speed is limited due to the amount of wrapper fibres generated. The speed of the rotor can be increased using smaller rotors because with smaller rotor diameter, less centrifugal forces are generated. This allows to further increase the rotary speed of the rotor and ultimately the productivity. For example, the centrifugal forces produced by a 30 mm rotor running at 120,000 rpm are approximately the same as that of 36 mm rotor running at 80,000 rpm. However this is also restricted by the fibre length because for a specific fibre length a specific rotor diameter has to be used.

In general all these factors related to wrapper fibre formation must be adjusted in such a way so that the wrapper fibre formation may be minimized. The wrapper fibres cannot be entirely removed otherwise the productivity will suffer a lot, however excessive wrapper fibre formation should be avoided as it degrades the quality and appearance of the yarn

Yarn Properties: The structure of open-end yarn is significantly different from ring-spun yarns. It may be seen from Exhibit 12 that fibres in a strand of open-end yarn are not as parallel as in ring yarn. This is especially true of surface fibres; many are simply wrapped about the strand of yarn randomly. However, the fibres near the centre of the yarn are more compact and contain more twist. The end result is a strand of yarn with a high twist, inner core of fibres surrounded by a sheath of wrapper fibres containing must less twist. This difference causes varying distribution of stress across the yarn from the axis to the surface and contributes to the uniqueness of open-end yarn. Some of the important properties of open-end yarn, compared with ring-spun yarn, are given below using the same fibre input into both:

- 10% to 30% weaker,
- higher elongation at break,
- more even,
- better abrasion resistance,
- less hairy,
- less shedding,
- bulkier, and
- larger yarn packages with no knots.

Open-end yarns are usually weaker than ring yarns despite the fact that generally higher twist multiples are used in open-end spinning. This is due primarily to a lesser degree of fibre parallelism in open-end yarns. However, this lack of orientation does tend to increase the elongation at break. Open-end yarns are more even than ring yarns due primarily to the suppression of drafting waves by the high doubling factor of fibre layering effect inside the rotor.

The abrasion resistance of open-end yarns is usually better than ring-spun yarns due in part to the higher twist multiples generally used and, to a large degree, to the yarn structure itself; the wrapper fibres carry relatively little load and, therefore, when they are abraded, the strength of the yarn is not seriously affected. An open-end yarn tends to have fewer hairs extending away from its surface. Whereas ring yarns may have as many as 90% of their fibre ends protruding from the yarn surface, an open-end yarn generally has 25% or less fibre ends protruding from its surface.

The reduction in surface hairs, plus the structure of the yarn itself, generally results in a reduction in the shedding propensities of open-end yarn.

An open-end yarn is usually somewhat larger in diameter than a corresponding ring yarn of the same count. This is due to a lack of parallel orientation of the individual cotton fibres, as is the case in ring-spun yarns. Consequently, open-end yarns are bulkier.

A typical ring-spinning frame produces a package containing such a small amount of yarn that during spooling or winding, many individual bobbins must be combined to produce a full package. To connect one bobbin to the next, it is necessary to knot or air splice the two yarn ends together. As a result, ring yarns may contain knots each 2-4 ounces of yarn. Most mills, however, have air splicer winders to eliminate knot defects. These automatic splicers produce quality yarn joints. Open-end machines produce packages that contain as much as 7 to 12 pounds of knot-free yarn and, as a result, the number of piece ups in open-end yarns is reduced dramatically and a much higher quality yarn results, especially when automatic piecing is used.

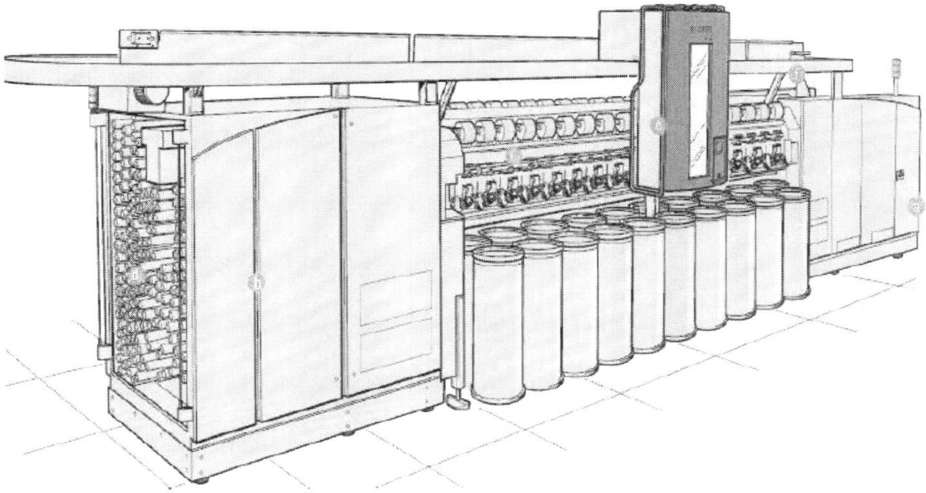

Figure 4.20 The schematic picture of a typical modern rotor spinning frame

A rotor Fibre length, longer the fibres being processed, larger rotor diameters are required otherwise the fibres will get damaged. So, when larger rotor diameters are used at high speeds for longer fibres, the amount of wrapping fibres will increase.

Friction between fibre & rotor groove, more the fiction generated between the two, more wrapping fibres will be generated.

Surface of navel, rougher the surface of navel, more wrapping fibres are generated and vice versa.

The increase in the rotor speed is limited due to the amount of wrapper fibres generated. The speed of the rotor can be increased using smaller rotors because with smaller rotor diameter, less centrifugal forces are generated. This allows to further increase the rotary speed of the rotor and ultimately the productivity. For example, the centrifugal forces produced by a 30 mm rotor running at 120,000 rpm are approximately the same as that of 36 mm rotor running at 80,000 rpm. However this is also restricted by the fibre length because for a specific fibre length a specific rotor diameter has to be used.

In general all these factors related to wrapper fibre formation must be adjusted in such a way so that the wrapper fibre formation may be minimized. The wrapper fibres cannot be entirely removed otherwise the productivity will suffer a lot, however excessive wrapper fibre formation should be avoided as it degrades the quality and appearance of the yarn

Yarn Properties: The structure of open-end yarn is significantly different from ring-spun yarns. It may be seen from Exhibit 12 that fibres in a strand of open-end yarn are not as parallel as in ring yarn. This is especially true of surface fibres; many are simply wrapped about the strand of yarn randomly. However, the fibres near the centre of the yarn are more compact and contain more twist. The end result is a strand of yarn with a high twist, inner core of fibres surrounded by a sheath of wrapper fibres containing must less twist. This difference causes varying distribution of stress across the yarn from the axis to the surface and contributes to the uniqueness of open-end yarn. Some of the important properties of open-end yarn, compared with ring-spun yarn, are given below using the same fibre input into both:

- 10% to 30% weaker,
- higher elongation at break,
- more even,
- better abrasion resistance,
- less hairy,
- less shedding,
- bulkier, and
- larger yarn packages with no knots.

Open-end yarns are usually weaker than ring yarns despite the fact that generally higher twist multiples are used in open-end spinning. This is due primarily to a lesser degree of fibre parallelism in open-end yarns. However, this lack of orientation does tend to increase the elongation at break. Open-end yarns are more even than ring yarns due primarily to the suppression of drafting waves by the high doubling factor of fibre layering effect inside the rotor.

The abrasion resistance of open-end yarns is usually better than ring-spun yarns due in part to the higher twist multiples generally used and, to a large degree, to the yarn structure itself; the wrapper fibres carry relatively little load and, therefore, when they are abraded, the strength of the yarn is not seriously affected. An open-end yarn tends to have fewer hairs extending away from its surface. Whereas ring yarns may have as many as 90% of their fibre ends protruding from the yarn surface, an open-end yarn generally has 25% or less fibre ends protruding from its surface.

The reduction in surface hairs, plus the structure of the yarn itself, generally results in a reduction in the shedding propensities of open-end yarn.

An open-end yarn is usually somewhat larger in diameter than a corresponding ring yarn of the same count. This is due to a lack of parallel orientation of the individual cotton fibres, as is the case in ring-spun yarns. Consequently, open-end yarns are bulkier.

A typical ring-spinning frame produces a package containing such a small amount of yarn that during spooling or winding, many individual bobbins must be combined to produce a full package. To connect one bobbin to the next, it is necessary to knot or air splice the two yarn ends together. As a result, ring yarns may contain knots each 2-4 ounces of yarn. Most mills, however, have air splicer winders to eliminate knot defects. These automatic splicers produce quality yarn joints. Open-end machines produce packages that contain as much as 7 to 12 pounds of knot-free yarn and, as a result, the number of piece ups in open-end yarns is reduced dramatically and a much higher quality yarn results, especially when automatic piecing is used.

A rotor spinning machine consists essentially of the following functional units (Fig. 4.20):

- Headstock (a) and tailstock (b) with central drives for rotors, feed, opening rollers and winding units; • the spinning and winding units (c), combined into sections; •

empty tube supply (d) with empty tube magazine and empty tube transport system (at the tailstock);

- ❖ 1 - 2 operating robots (e) on each side of the machine for cleaning, piecing and package change; • package conveyor belt for transporting the full cross-wound packages to the end of the machine (f);

In summary, when cottons are selected for open-end spinning, they ideally should have low levels of non-lint, and the fibres should be as mature and strong as possible. The appearance of open-end yarn in fabric (especially knitted) is more uniform and, therefore, desirable for most applications.

1. Rotor spinning has been characterized from the outset by incomparably higher,production potential than ring spinning.
2. This potential has been steadily increased by the continuous rise in rotor and winding speeds.
3. Rotor-spun yarns have therefore always been successful where they could be manufactured more cheaply than ring-spun yarns and proved suitable for the range of application in question.
4. Rotor spinning combines two process stages – spinning and winding – in a single machine.

The potential of rotor spinning

From the multitude of spinning processes developed in recent decades, e.g. Bobtex, Repco, Twilo, friction, Air-jet and wrap spinning, only rotor spinning and – with reservations – Airjet spinning have established themselves successfully on the market.

Essentially, both the technological and the economic potential of rotor spinning are the decisive factors in the success of this spinning process.

Although many developments are carried out in rotor spinning even now rotor spinning cannot produce finer yarns otherwise easily possible to produce with ring spinning The presence of wrapping fibres in the rotor spun yarn gives it a harsher and hairier look.

The rotor spun yarns have high pilling property as compared to ring spun yarn.

The strength of the rotor spun yarn is less than that of the ring spun yarn (about 70% of the ring spun yarn).The rotor spun yarn causes more wear of knitting needles and loom components as compared to the ring spun yarn. The rotor spinning process is more complicated as compared to ring spinning and the know-how of operation and maintenance is less as compared to ring spinning which is widely accepted and universal form of spinning.

The false twist effect generated between the draw-off nozzle and the yarn unrolling from it has Z twist between the draw-off nozzle and the rotor groove and S twist between the draw-off nozzle and the nip of the take-off shaft and the pressure roller. At this nip the false twist effect has again reached its zero point and the yarn body has only the preset genuine Z twist.

In mill operations and the overwhelming majority of applications, rotor cleaning is performed automatically at each piecing operation at the spinning position, i.e. at each end down, each quality stop and each package change.

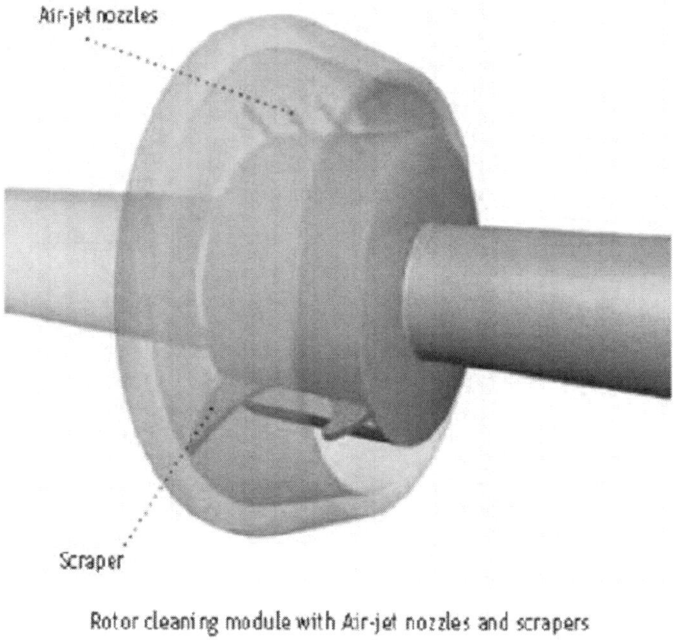

Figure 4.21 Rotor cleaning module

Since a clean rotor groove is the precondition for both successful spinning start-up and high piecing quality, on modern systems the rotor groove is cleaned by means of a rotating cleaning head.

The cleaning head cleans the rotor groove with 2 scrapers, while 3 air jets clean the rotor slip wall and the groove. The cleanliness of the rotor groove and the rotor that is required for trouble-free spinning operations is adequately assured by the frequency of the piecing process and the resulting cleaning intervals.

Essentially, two systems are used to clean the rotors: pneumatic cleaning by means of compressed air and mechanical cleaning by means of scrapers. Both systems are also used in combination (see Fig. 4.21). During rotor cleaning the surface of the draw-off nozzles and the draw-off tube are also cleaned. Further modules clean the nozzle surface and the draw-off tube attached to it, either mechanically with a brush, or pneumatically with an Air-jet.

We distinguish between two different rotor bearing systems:

❖ Direct rotor bearing (Fig. 4.22), in which tangentially driven rotor shaft (a) is encased in ball bearing housing (b). The ball bearing rotates at the same speed (rpm) as the rotor shaft driven by the tangential belt. This bearing principle limits rotor speeds to approx. 110 000 rpm. Although direct bearings would be ideal, individual motors have also been unable to establish themselves for this rotor drive, on cost grounds.

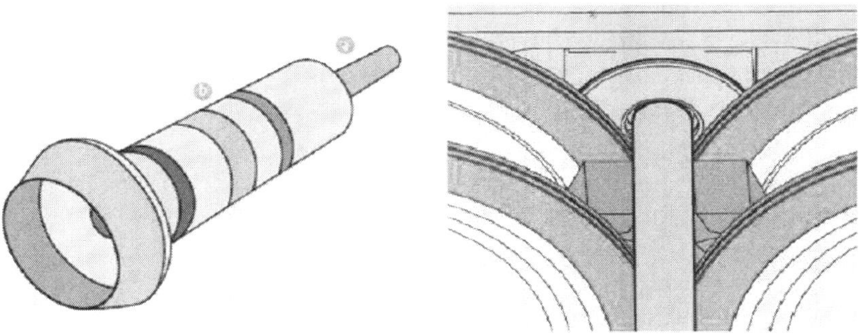

- Direct rotor bearing, with rotor shaft (a) encased in ball bearing housing (b)

- Support disc bearing (Twindisc bearing) with rotor fitted

Figure 4.22 Direct rotor bearing (encased in ball bearing and support disc bearing)

Indirect rotor bearing, in which the rotor shaft, also driven tangentially, runs on two pairs of supporting discs arranged side by side (see Fig. 4.23). With the support disc bearing the rotor speed is reduced at a ratio of 1:8 to 1:10 relative to the bearing of the supporting discs, depending on the diameter of the discs, so that these bearings run at speeds of only 16 000 to a maximum of 20 000 rpm (depending on the diameter of the supporting discs), even at rotor speeds of 160 000 rpm.

For one thing, this bearing system permits much higher rotor speeds than direct bearings, and at the same time the service life of indirect bearing systems is significantly higher than that of directly driven bearing systems. High-performance rotor spinning machines operating at speeds of up to 160 000 rpm are therefore operated only with indirect rotor bearing. As already stated, with both bearing systems the rotors are driven by a tangential belt on each side of the machine, the speed of which can be adjusted either by stepped speed pulleys or steplessly by means of an inverter drive.

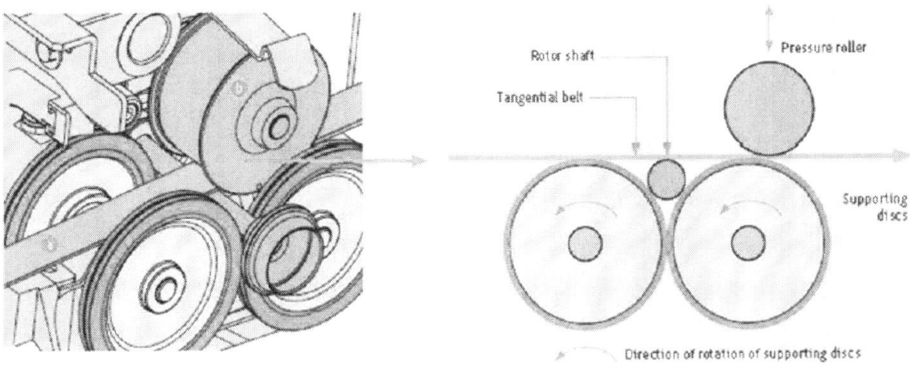

Support-disc bearing (Twindisc bearing) with pressure roller (b) for tangential belt (a)

Figure 4.23 Tangential belt drive of the rotor

Tangential belt (a) is engaged with the rotor shafts via pressure rollers (b) to drive the rotors (see Fig. 4.20). If a spinning position is stopped and the rotor cover opened, the tangential belt is disengaged at this spinning position by raising the pressure roller and the rotor shaft is brought to a standstill by a brake positioned between the supporting discs. Since the rotor is held in position only by the light pressure of the tangential belt on the support-disc coatings, it can be removed very easily without the use of tools for replacement or examination and refitting.

Whereas both the tangential and the axial position of the rotor are defined by the fixed ball bearing housing in the case of direct rotor drive, the rotor on support-disc bearings also has to be fixed in position in the axial direction. The rotor is fixed in position axially by slightly crossing the pair of supporting discs, so that the rotor is pressed backward with some force (toward the spinning beam). Various bearing systems are available for absorbing this backward axial pressure:

Magnetic bearings (see Fig. 4.24). The end of the rotor shaft is fixed in position without contact in a magnetic field created by annular magnets. Accurate radial positioning of the rotor shaft is the precondition for the functioning of this system, which as far as is known to date has no speed limitations.

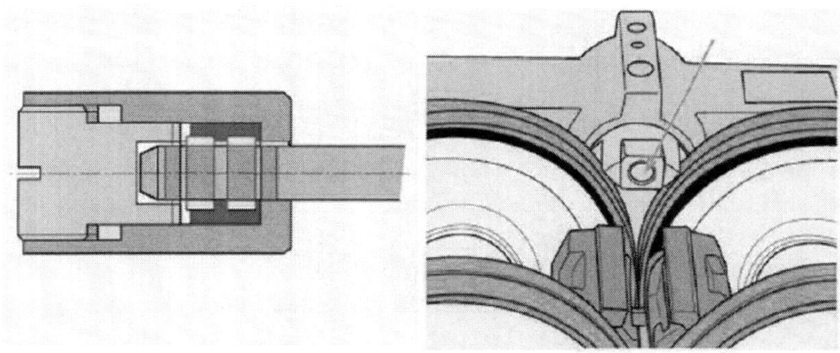

- Axial rotor bearing with magnetic bearing   Positioning the magnetic bearing

Figure 4.24 Axial rotor bearing with magnetic bearing

EC bearings (Fig. 4.25). The end of the rotor shaft runs (in contrast to the oil bearing) on a steel ball embedded in grease. The housing is sealed, grease cannot escape, and the bearing is largely maintenance-free.

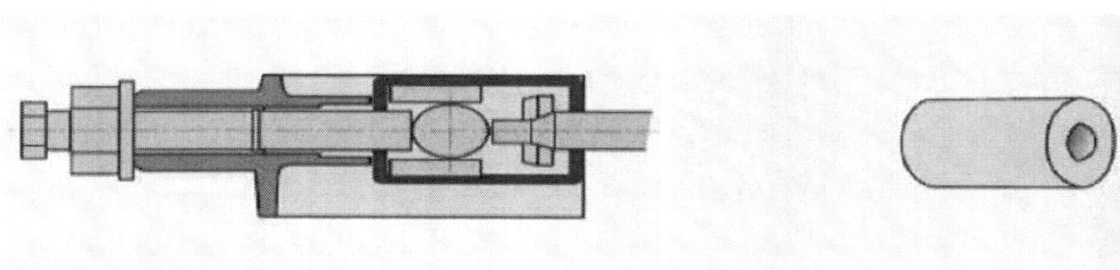

Axial rotor bearing with EC bearing   Sealed grease cup of the EC bearing

Figure 4.25 Axial rotor bearing with EC bearing

AERO bearings (Fig.4.26). In this bearing system an air cushion provides axial support for the rotor. This air cushion is provided by a compressed air supply of 6 bar to each spinning position.

This system requires neither oil nor grease, sticky deposits are avoided, and in the immediate vicinity of the air cushion the permanent current of air ensures continuous cleaning (selfcleaning effect). Other advantages of this system are low maintenance effort and spare parts consumption. The accurate, level surface of the end of the rotor shaft is the precondition for trouble-free operation.

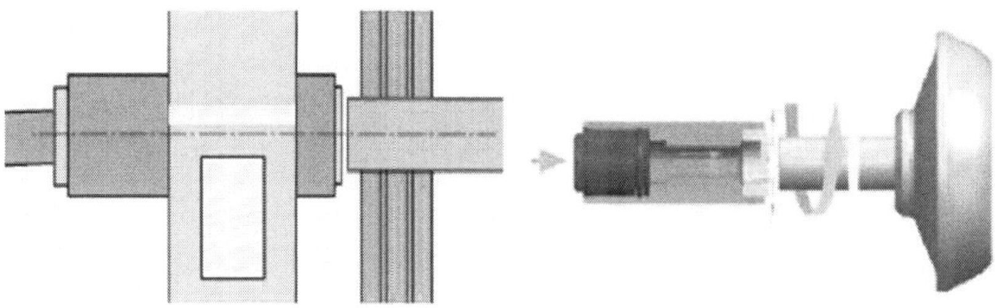

Figure 4.26 Axial rotor bearing with magnetic bearing

This rolling-off temporarily inserts additional twist into the yarn (contrary to the direction of twist of the yarn), thus creating the false-twist effect required for spinning stability, which can be up to 60 % of the set yarn twist. The greater the false-twist effect, the higher the spinning tension.

Yarn take-off

The yarn is taken from the rotor by the delivery shaft and pressure roller, diverted virtually at right angles in the process by draw-off nozzle projecting into the rotor and guided out by drawoff tube immediately following this. However, as has already been said, the draw-off nozzle is far more than a mere guide device. At take-off the yarn continuously rolls off on the surface of the draw-off nozzle due to the rotation of the rotor. Almost all rotor spinning machines nowadays produce packages with a traverse of 150 mm (6″), which results in the following package formats, depending on the winding unit of the different types of machine:

- ❖ Cylindrical packages: max. diameter 350 mm; max. package weight up to 6 kg;
- ❖ conical packages (2° - 4°51'): max. diameter 280 mm; package weight depends on package density.

The high package weights reduce handling costs in the spinning mill and downstream processing as well as capital costs for empty tubes.

Compensation of winding tension

The traversing motion of the yarn depending on stroke and winding helix requires compensation of the winding tension for homogeneous package density. A thread guide is arranged to effect crosswise laying of the yarn in the package by means of its to-and-fro movement.

A compensation bow (Fig. 4.27 a) and the yarn tension bar (Fig. 4.24, b) are needed to even out path-length variations which arise because the length of yarn between the take-off rollers and the right- or left-hand edge of the package is greater than the length between those rollers and the middle of the package.

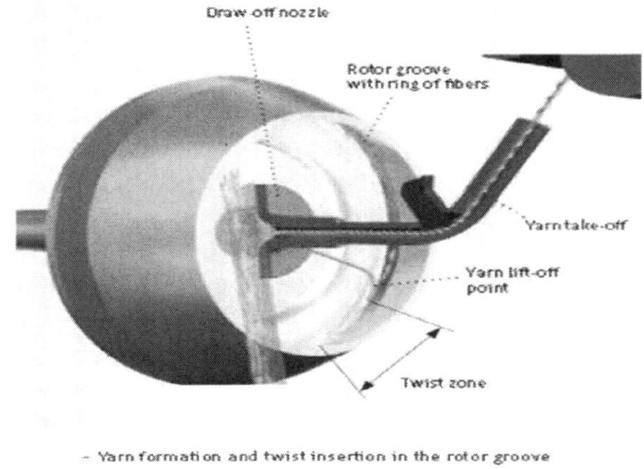

Figure 4.26.Yarn formation and twist insertion in the rotor groove

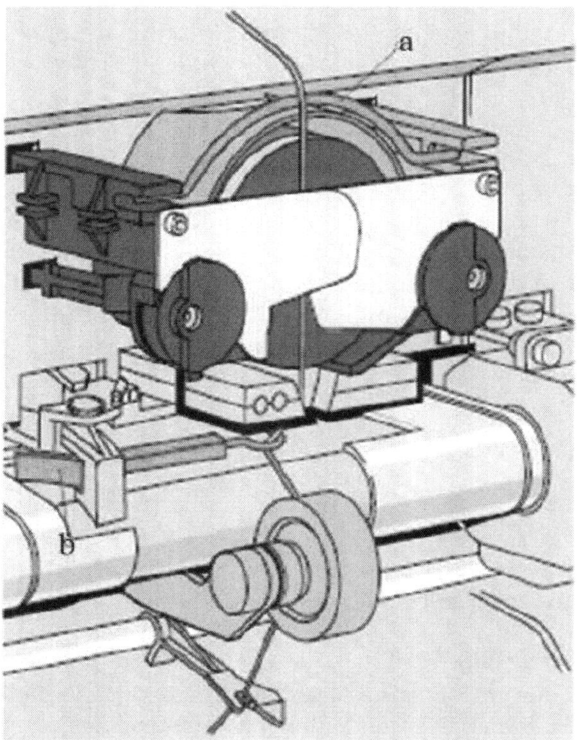

Figure 4.27 Yarn winding on rotor spinning machine

As already mentioned, the density (γ) of the package depends not only upon the winding angle but also on:

- ❖ the (adjustable) winding tension;
- ❖ the (adjustable) contact pressure of the package on the winding roll; and • the yarn count.

A finer yarn always gives a higher package density, which can be calculated according to the well-known physics equations:

density (γ) = mass/volume,

γ = *yarn net mass (g)/ yarn volume (cm3)*

The settings for draft (ratio n feed cylinder/ n delivery cylinder), yarn twist (ratio n rotor speed /n delivery cylinder) and winding tension (ratio N delivery cylinder/n winding shaft) are made via the drives for the feed cylinder, the take-off rollers and the winding shaft. The interaction of the drives for draft, twist and winding tension is shown schematically in Fig. 4.28. Settings are made either via infinitely adjustable inverter drives or conventionally via change gears.

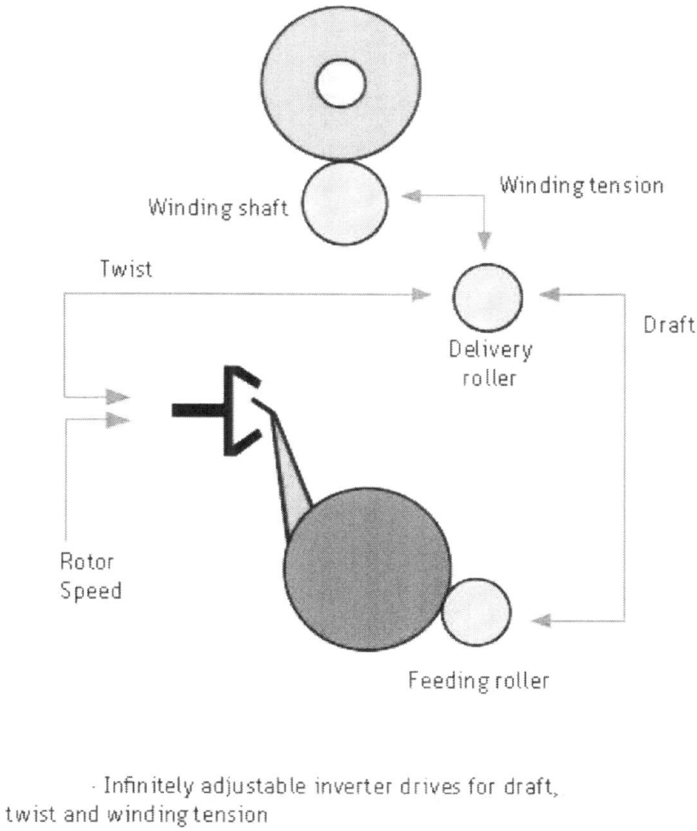

Figure 4.28 Schematic of setting points

Suction system machine

The main fan (Fig. 4.29) sucks the air from each spinning position through negative pressure duct (2) and filter housing (3), thus creating a negative pressure of approx. 60 - 85 hPa at the rotor housing of the spinning box. Trash, dust and fibre fragments carried by the air current are collected by a filter layer (4) in the filter housing. The layer of fibres, trash and dust is held on the filter by the current of air. However, as the filter becomes increasingly full, the negative pressure inevitably weakens. If the negative pressure then falls below the adjustable limit value (alarm level), the current of air is automatically diverted briefly through a bypass (5). The layer of material on the filter is now no longer held and drops onto the base of the filter housing (6).

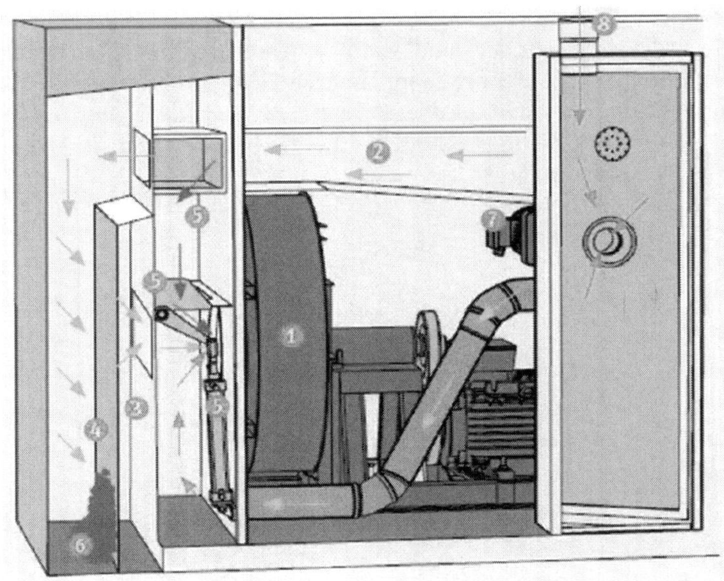

Figure 4.29 Schematic illustration of suction system

Quality control systems

The fact that rotor-spun yarns contain significantly fewer yarn defects than ring-spun yarns has made a major contribution to the success of the rotor spinning system. A comparison of Uster Statistics shows that the numbers of thick places, thin places and neps are significantly below the level of ring-spun yarns, even at delivery speeds that are up to 10 times higher.

And the finer the yarn counts, the greater the differences. The reasons for this are back-doubling in the rotor (which balances variations in mass) and fibre guidance and monitoring without a cylinder drafting system. Furthermore, a package of rotor-spun yarn contains only a fraction of the yarn joints (piecings) compared with a cross-wound package of ring-spun yarn. A 4 - 5 kg cross-wound package in the rotor spinning mill contains no more than 3 - 5 spinning- related piecings at normal ends down rates. However, a 3 kg cross-wound package of ring-spun yarn produced on the winder already contains some 30 - 40 piecings due to system-related cop joints plus a certain number of additional piecings due to cleared yarn defects. This very soon adds up to more than 50 piecings (splices or knots per package).

Leading global suppliers of quality control systems (e.g., Uster Technology with the Uster Quantum Clearer2® and Barco with the Barco Profile) employ different measuring systems in some cases, but offer a largely comparable range of performance:

1. detecting, counting and clearing disturbing yarn defects in accordance with adjustable clearing limits;
2. counting uncleared (non-disturbing) yarn defects in defect classes;
3. detecting and eliminating extraneous substances;
4. measuring the main physical textile yarn attributes: yarn irregularity, imperfections and Classimat values (not yarn tenacity and elongation).

	Capacitive principle	Optical principle
Property • Yarn	Corresponds to the mass of the yarn, or number of fibers	Corresponds to the diameter of the yarn, the visual impression
Influence • Fiber	Yarn, contains electrical conductive fibers cannot be processed	All fibers
• Colored	No	Dark yarns possibly require other settings
• Fiber	No	No
• Humidity	Variations in the humidity can have an influence	No influence: very dry yarns exhibit a higher hairiness - larger diameter - unjustified stops

Properties of the measuring principle

Gives the details of sensitivity of the measurement systems

Yarn		Capacitive principle	Optical principle
Regular yarn		0 % base value	0 % base value
Thick place with double cross-section		Increase of cross-section: +100 %	Increase of diameter: +42 %
Thin place with half cross-section		Decrease of diameter: -50 %	Decrease of diameter: -29 %

Sensitivity of the measuring principle

Figure 4.30 Sensitivity of measuring principle

Machine automation in rotor spinning

Systems for automating the operation of rotor spinning machines have been integral parts of high-performance rotor spinning machines for some years. Automated systems have been developed for all manual operations in several stages:

1. automatic gripping and introduction of the sliver end from a new can into the spinning box (implemented only in certain cases to date);
2. automatic cleaning of rotor, draw-off nozzle and draw-off tube after ends down, quality stops or package changes;
3. automatic piecing (start-up) after ends down, quality stops or package changes;
4. automatic removal of full packages upon reaching the preset yarn length, and replacement with empty tubes;
5. automatic feeding of empty tubes to the operating robot for package change; programmable batch phase-out/batch change;
6. automatic deposit of removed packages at the end of the machine;
7. automatic or semi-automatic filter cleaning.

Machine automation is represented in practice by **two different concepts**:

❖ **Integrated automation**, in which all operating functions (rotor cleaning, repairing ends down, and package change) are combined (integrated) in a single robot. Package changing and the subsequent re-start of the spinning position occur as a single process.

❖ **Automation by means of units operating separately**, with the operating functions of spinning start-up (after ends down or package changes) being performed by a **piecing robot**, and the transport of starter bobbins (instead of empty tubes) and package change by a **second robot**. There is no system-imposed link between robots which operate separately and the use of starter bobbins, but the greater technical complexity this concept entails in connection with the pre-wound starter bobbin (additional starter bobbin unit, starter bobbin transport, etc.) is system-imposed.

Machines with two operating robots

Two operating robots (on for each side of the machine) are usually adequate for serving the spinning positions efficiently on the machine lengths of 240 to 280 spinning positions that are customary nowadays. Each robot serves one side of the machine, and when one robot is being serviced the second robot can be programmed also to serve the other side of the machine (Fig. 4.31). In this case the robot transfers from one side of the machine to the other via a loop on the headstock. The downtime due to stationary spinning positions is therefore reduced by half.

Using 4 robots on machines of this length results in a significant improvement in efficiency only in extreme spinning conditions.

Operating robot on a modern high performance rotor spinning machine

Figure 4.31 Operating robot

In the context of transport automation the rectangular can has been launched on the market and has enabled can change on the rotor spinning machine to be automated (Fig. 4.31a). Rectangular cans enable the available space to be allocated clearly in terms of the spinning can and the spinning position (only one can stands under each position). This layout of spinning can and spinning position in relation to the gauge was one of the essential preconditions for simple, automated can change on the rotor spinning machine.

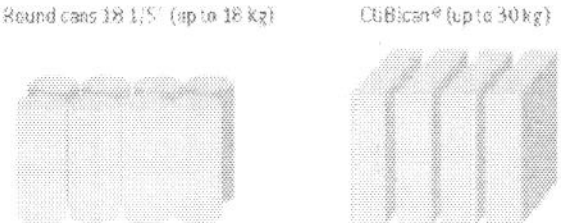

– Can formats in the rotor spinning mill.
Rectangular cans for economical transport automation

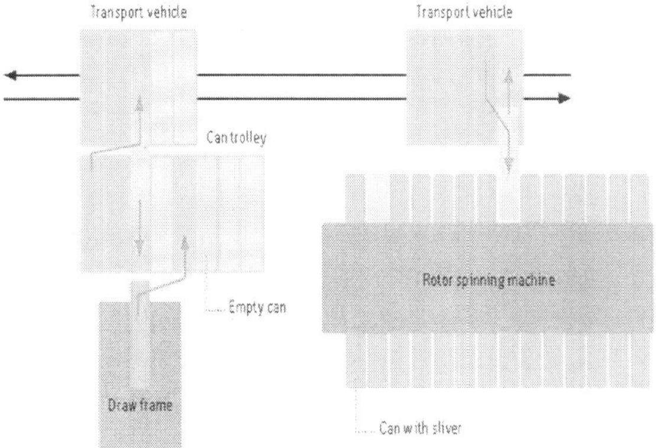

automatic can transport between draw frame and rotor spinning machine

Figure 4.31a Automatic Can Transport arrangement

Rotor

The rotor is the main spinning element of the rotor spinning machine. Yarn quality, yarn character, operating performance, productivity, etc., all depend chiefly on the rotor. The most important parameters of the rotor that exert influence are (see Fig. 4.31):

- the inclination of the rotor wall (a);
- the coefficient of friction between the fibres and the surface conditions of the rotor wall (b);
- the design and the positioning of the rotor groove (c);
- rotor groove diameter (d) and rotor speed.

Out of the many types of rotors available, the spinner has to select the one that is best suited to the raw material, yarn product, and spinning conditions. Rotors are replaceable elements in all rotor spinning machines.

The rotor, consists of rotor shaft (a) with wear protection in some cases, rotor cup (b) with rotor groove © and rotor wall (d). The wall inclination is necessary so that fibres emerging from the feed tube and passing to the wall can slide downward. Depending upon the material and area of use, the angle of the rotor wall to the vertical ranges between 12° and 50°. This angle is dependent upon the make but will in all cases be smaller, the higher the rotation speed for which the rotor is designed. At the internal periphery in the lower region of the rotor cup, there is usually a groove that varies in width. This groove serves to collect fibres.

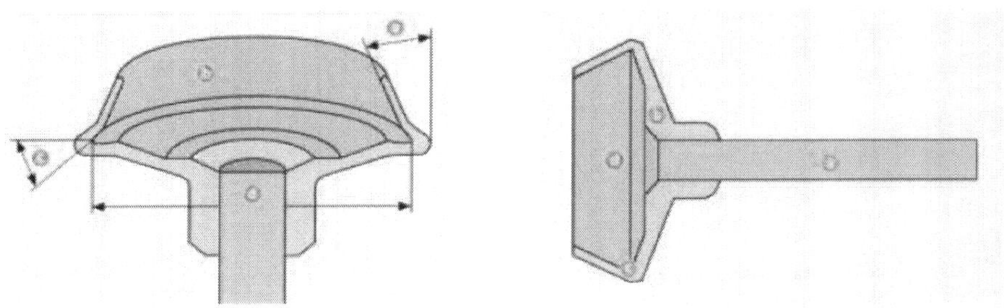

4.32 Important Rotor Parameters

However, the surface structure of the rotor wall changes depending on the type of treatment (boron or diamond coating), and thus also its influence – which should not be underestimated – on both yarn quality and spinning stability and the tendency for deposits to form in the rotor groove. The best possible compromise between long service life of the rotor, good yarn values and stable spinning conditions is achieved with the combined boron/diamond coating. The rotor is a part subject to wear and must therefore be replaced periodically. Wear mainly affects the groove.

The configuration of the rotor groove determines whether the yarn is bulky or compact, hairy or lean, and whether the yarn quality is excellent or only adequate and the spinning stability low or high. The groove also affects the extent to which dust and dirt tend to accumulate in the rotor. Depending upon the raw material used, the desired yarn characteristics and yarn values, different groove designs are used in practice.

Wide grooves produce a soft, bulky yarn with rather low strength, while narrow grooves produce a compact, strong yarn with low hairiness.

Wide grooves are therefore used in the production of yarns for knitted fabrics, homespun-type fabrics and coarse articles; narrow grooves are used for yarns required for the production of stronger fabrics with a smooth appearance.

A fairly narrow groove is in most widespread use in classical short staple mills. The tendency to form moiré effects is also greater with the narrower groove, because fairly large dirt particles can jam in the groove.

A speed range in which the given rotor produces the optimum results, in terms of technology as well as spinning stability and energy consumption, can be assigned to each rotor diameter. The speed ranges overlap between rotor diameters, with the energy consumption of the smaller rotor diameter being more favorable at the same rotor speed.

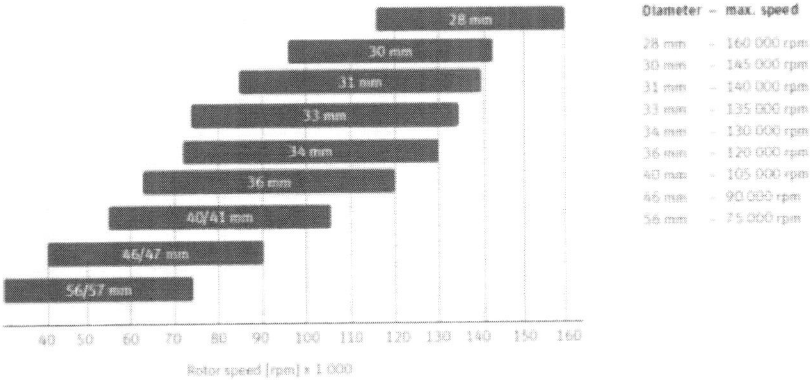

Figure 4.33 Speed range and maximum rotor speed for a given rotor diameter

The rotor diameter should in any event be large enough to permit fibre formation in the groove without technological disadvantages. A certain amount of space is needed for the fibre mass, i.e., larger rotor diameters have to be used for coarser yarns and vice versa. A relationship – (not very close) – also exists between fibre length and rotor diameter.

As a rule of thumb, rotor diameter should not exceed 1.2 times staple length, otherwise fibre integration in the rotor groove is disturbed.

Fig. 4.34 describes the configuration and the properties of the different rotor and groove shapes.

In principle:

- ❖ Narrow groove angles and small groove radii (T and K rotors) are suitable for all raw materials and are used to manufacture smooth weaving yarns with good regularity and high yarn tenacity.
- ❖ Narrow groove angles with large groove radii (G rotors) are also suitable for all raw materials and are preferably used for bulky knitting yarns.

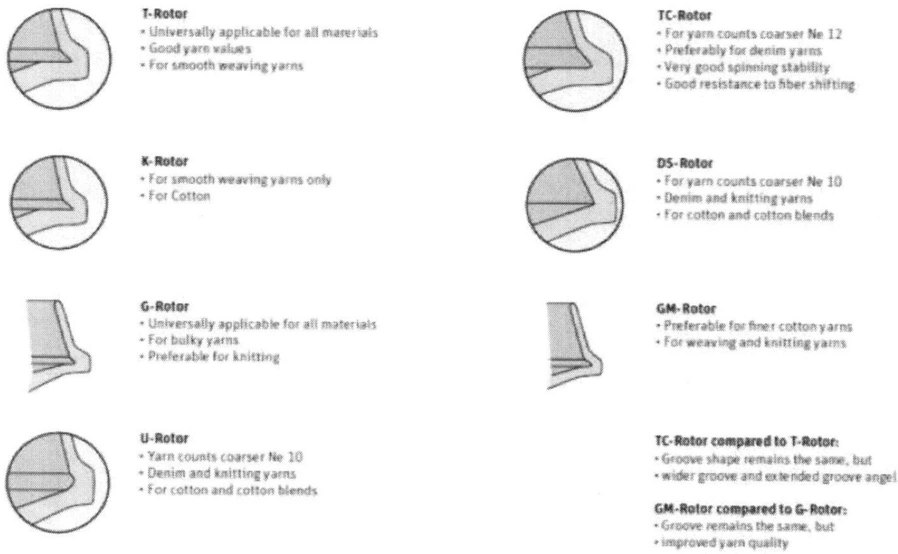

Figure 4.31 Configuration of rotor/groove shapes and yarn properties

Rotors with wide groove angles (U and DS rotors) are suitable for bulky knitting and denim yarns in cotton and its blends with man-made fibres. The different groove shapes and groove radii are chosen according to the type of denim yarn (weft or warp yarn, rope or beam dyeing, etc.).

- ❖ The TC rotor is outstandingly suitable for manufacturing high-quality denim yarns and at the same time is characterized by excellent running properties. Compared to the T rotor, groove angle and groove radius are larger, but the groove shape has been retained. Especially shiftingresistant yarns are produced when processing man-made fibres and viscose with the TC rotor.
- ❖ The GM rotor can be used very flexibly in the fine count cotton yarn sector, for both weaving and knitting.

Compared to the G rotor, groove angle and groove radius are larger, but the groove shape has been retained.

Draw-off nozzles are made of either ceramics or steel. Drawoff nozzles usually consist of two parts, a wear-resistant ceramic nozzle head and a metal nozzle holder (Fig. 4.35).

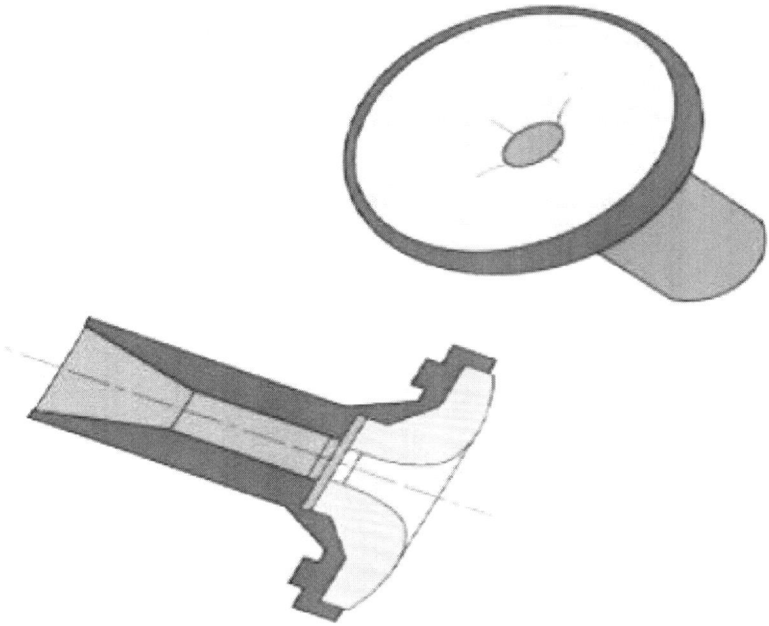

Figure 4.35 Draw-off nozzles with ceramic nozzle head and metal nozzle holder

Ceramic or metal draw-off nozzles are also in use in which nozzle head and nozzle holder are produced in one piece. There are no technological differences, except that solid ceramic drawoff nozzles feature very low heat dissipation (ceramics are used as insulators in electrical installations) and can therefore hardly be considered for processing manmade fibres. If the rotor groove makes a crucial contribution to yarn quality and bulk in technological terms, the structure and design of the nozzle surface exert a decisive influence on surface structure and hairiness. The further the draw-off nozzle protrudes into the rotor, the larger the yarn's angle of wrap at take-off, the more false twist is created and the longer the binding zone in the rotor groove. In some cases this can help to increase yarn tenacity.

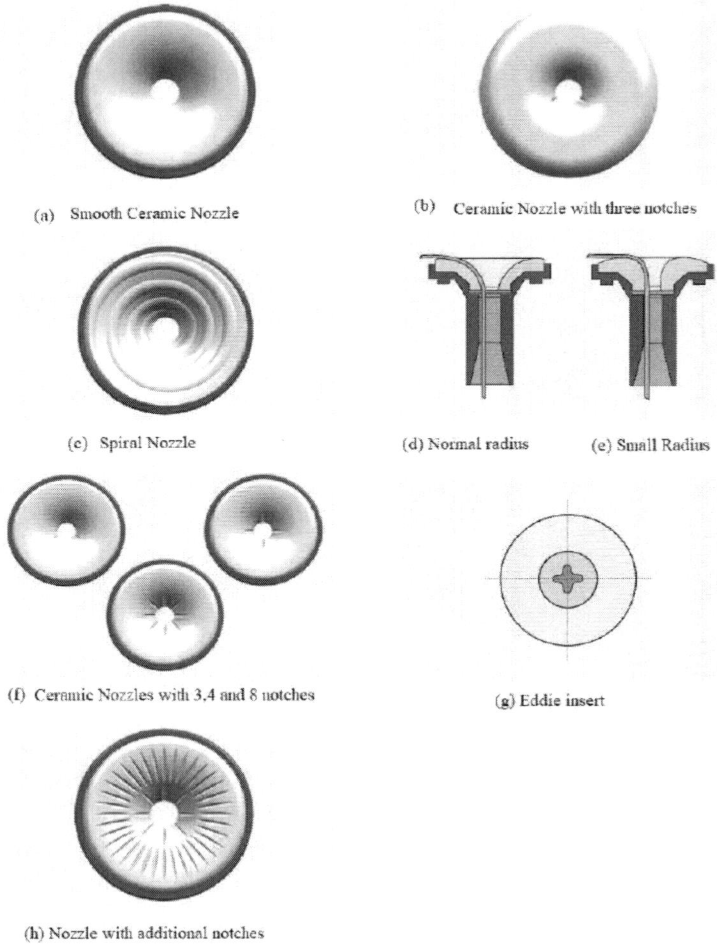

Figure 4.36 Different types of nozzles

Nozzles with a smooth surface (Fig.4.36 a) are suitable for producing smooth warp yarns with low hairiness. This type of nozzle is rarely used, since very high yarn twist has to be imparted due to the low level of false twist created. Yarn values are not better than with other nozzle types in every case.

- ❖ Nozzles with a spiral surface (Fig.4.36 b) are ideally suitable for compact and fi ne warp yarns in 100 % cotton with low hairiness and good yarn values. High spinning stability.
- ❖ Nozzles with 3, 4, 6, 8 or more notches (Fig. 4.36) are universally applicable both for cotton and also for man-made fibres and their blends

The greater the angle of diversion and the higher the friction caused by the additional ceramic inserts, the greater the twist retention, the more intensive the twist propagation into the rotor groove and the higher the spinning stability. This favors the manufacture of especially softtwisted knitting yarns, since the high twist retention enables low twist multipliers to be set without adversely affecting spinning stability.

Figure 4.37 The major Spinning elements for rotor spinning machines, opening rollers (rear left), rotors (rear right), draw-off nozzles (front), channel plates (center)

4.8 The processing stages

In rotor spinning, both the characteristics of the raw material and the manner in which this material is prepared on the mill's preparatory equipment are important; as well as significant. The machines to be selected and the processing lines that must be adapted to the type of raw material have to be planned well. Currently, the processing lines shown in Fig. 4.37 are most commonly used.

A third draw frame passage is not even necessary when cotton is blended with synthetic fibres in sliver form, because the back-doubling in the rotor leads to a high degree of fibre/fibre transverse doubling.

Blowroom

Since rotor spinning reacts less critically to short fibres than ring or Air-jet spinning, the main task of blowroom machinery is the efficient removal of trash and dust. The blowroom line can therefore be kept very short, but calls for very effective cleaning and opening units

Cards

The card usually has to reduce the dirt content to less than 0.1 - 0.2 % and also to remove part of the dust. The card is already capable of removing dust adhering to the fibres because significant fibre/metal friction arises here, and the dust is rubbed off. With regard to dust removal, the blowroom, carding room and draw frames are each expected to remove about onethird of the dust. Web crushing at the delivery of the card often brings about a significant improvement in the cleaning effect for cotton with medium to high dirt content.

When the carded sliver is processed directly on the rotor spinning machine (Fig. 4.36) the card must be equipped with a levelling device or a card with a draw frame module used.

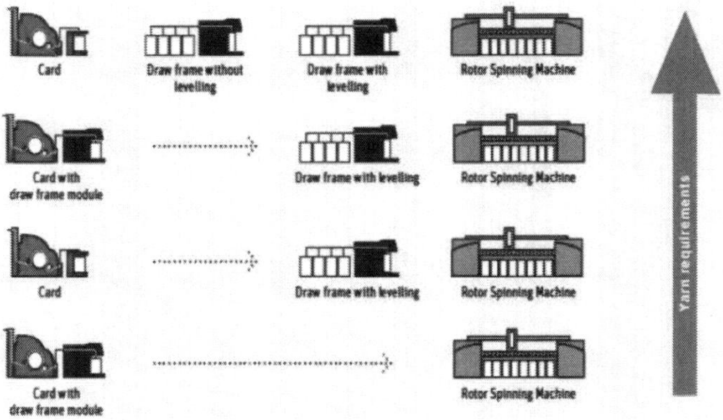

Fig. 4.38 – Rotor spinning systems with different sliver preparation passage routes

4.9 Economics of Rotor Spinning

Any new spinning process launched on the market can only be successful if it fulfills certain criteria for economic benefits and can claim advantages over an established spinning system in at least one of these criteria, such as:

1. higher quality of the product manufactured;
2. higher productivity of the system as a whole;
3. lower costs of the production process in relation to the quantity produced (labor, energy, capital);
4. greater flexibility of the process, i.e. a wide range of yarn products can be manufactured or a wider range of raw materials can be used.

The general comparison of ring spinning and rotor spinning costs are given in figure

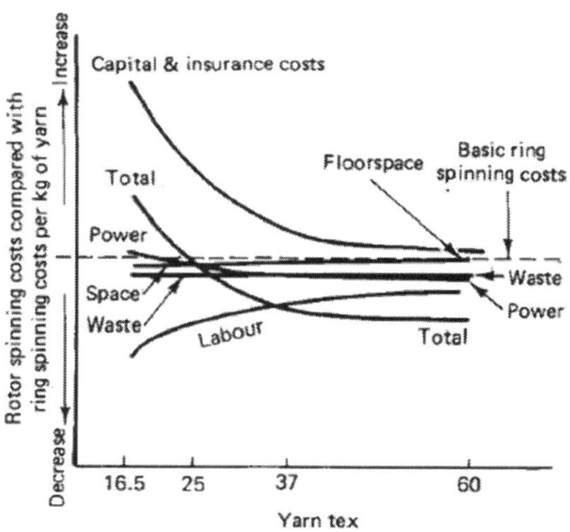

General comparison of ring spinning and rotor spinning costs

The stress-strain behaviour of rotor spun yarns is largely identical to that of the ring spun yarns.

The same can be seen from the following table.

The stress-strain behavior of the rotor-spun yarn is largely identical to that of a ring-spun yarn.

Rotor-spun yarn compared to ring-spun yarn		
tenacity cN/tex	lower	-
CV% cN/tex	lower	+
elongation at break %	higher	+
irregularity CV%	lower	+
imperfections / 1 000 m	much lower	++
yarn bulk	higher	+
abrasion resistance	higher	+
stiffness	higher	1)
hand	harder 2)	-
surface	rougher	1)
hairiness	lower	1)
luster	duller	-

1) Positive or negative, depending on the end product

2) Virtually the same as ring-spun yarn in the end product after finishing

All rotor spinning machines are designed to spin yarns with Z twist. Z twist is the customary direction of twist used in practice. Manufacturing yarns with S twist would imply redesigning the rotor drive, sliver feed into the spinning box and fibre feed to the rotor.

As far different varieties of yarns that can be produced, the rotor spinning system is a less flexible system as compared to ring spinning system.

Rotor spinning system cannot be employed for woollen and worsted fibres.

Chapter 5

Friction Spinning System (DREF)

5.1 Introduction

In the last twenty-five years, the textile industry has witnessed some exciting innovations in spinning machines. The machinery manufacturers had been looking for alternatives to ring spinning. When BD 200 rotor frame was marketed in 1967, the momentum for the search of alternative technologies really picked up. Rotor spinning technology is being adopted gradually in India to spin cotton counts up to 24 Ne. On the other hand the air-jet technology, which is new at present, is claimed to be suitable for spinning medium and finer counts. The electrostatic open-end spinner that was exhibited at the 1971 ITMA exhibition aroused widespread interest, and many people have continued to expect the commercial machine. However, they continue to wait. After early disappointments, and the virtual dismissal of Vortex Spinning techniques, the Monsanto Company had patented a fluid Vortex method for the production of composite yarns consisting of very short high modulus refractory fibres and man-made fibres to act as carriers. The inventors of Bobtex processes of yarn production have given their attention to open-end spinning and patented two processes, both of which are intended to give a twisted strand of continuous filaments, and although they are described by their inventors as open-end or break spinning techniques, it appears that neither of them incorporates the essential feature of a true open-end system. Friction spinning is defined as an 'open-end' spinning method (or an OE spinning method), in which the yarn formation takes place in the yarn forming zone consisting of two friction rollers with the aid of frictional forces. In traditional spinning systems the fibre supply is first formed into a strand of fibres, almost equal to the final count of the yarn, and then twist is inserted to form the yarn. With OE friction spinning the fibre supply must be opened completely into individual fibres and reassembled at the nip line of the two friction rollers to form a twisted strand of fibres, i.e. the yarn, each fibre being twisted onto the forming strand by the friction rollers as soon as it reaches the nip line. While the production speeds are limited with both ring and rotor spinning, the friction spinning system has a low spinning tension and is therefore suitable for the production of yarn at high speeds of up to 500 metres per minute.

Owing to the absence of positive control over the assembly of fibres, much slippage occurred between the fibre assembly and the perforated roller, which reduced the twist efficiency. Hence this particular design could not be commercialized. In order to reduce slippage and improve the twist efficiency, the concept of enclosing the fibre assembly between two perforated friction rollers was devised, and formed the basis for the commercial development of two designs of DREF spinning machines, known as DREF-2 and DREF-3. The DREF-2 machine was demonstrated at the 1975 ITMA exhibition and then commercialized in 1977. The DREF-3 machine was introduced on the market at the end of 1981.

The DREF-3 machine was developed basically to improve the yarn quality, extend the yarn count towards the finer end of the count range (up to 33 Tex), and produce multicomponent yarns. As an improvement in respect of the vertical fibre-feeding arrangement that exists in DREF-2 and DREF-3 machines, the DREF-5 friction-spinning machine was developed jointly by Schlafhorst, Suessen, and Fehrer. The DREF-2000 and DRE F-3000 are the latest development in friction spinning. Since 1967, many friction-spinning systems have also been patented by various companies and individuals. Although these machines have different modes of operation, their basic principle is similar. The first patented friction-spinning system was granted to Platt-Saco-Lowell in 1967. The company much later developed and introduced their friction spinning machine, called the Master-spinner, at the 1983 ITMA exhibition in Milan. An alternative frictionspinning system was also developed by the company Ernst Fehrer and exhibited in 1973 under the name of DREF-1. With this system, the opened fibres were made to fall on a vacuum slot of a single perforated cylindrical roller, the rotation of Friction spinning belongs to the family of open-end spinning. Most patents related to friction spinning were filed in the 1970s and 1980s, many of which were from Dr Ernst Fehrer in Austria. Today friction spinning is almost synonymous with the term DREF (**Dr E**rnst **F**ehrer). It has been used to produce yarns usually much coarser than ring and rotor spun yarns at much higher production rate, and the yarns have been largely used for domestic and industrial applications. This chapter discusses the principle of friction spinning in general, followed by a discussion of the DREF 2 and DREF 3 friction spinning systems.

5.2 Open-end spinning processes- The basic principle of yarn formation

In all other spinning processes, an uninterrupted stream of fibres proceeds continuously, but with gradual attenuation, from the feedstock to the take-up package. In open-end spinning, this flow of fibres is interrupted, the fibre strand being opened into individual fibres at a predetermined position, usually by means of an opening roller, followed by airborne fibre transport.

This interruption or break in the fibre flow is physically achieved by increasing fibre speed locally to very high levels (up to 100 m/s), so that – according to the equation of continuity the number of fibres in the cross-section drops to such low values that the fibres lose contact with each other.

A constant stream of separated, individual fibres is allowed to flow to a rotating yarn end. The brush-like, open yarn end grasps the fibres brought into contact with it and continuously binds them into a yarn with the aid of the continual rolling movement. This is schematically illustrated in figure 5.1

The continuously formed yarn has only to be withdrawn and taken up onto a cross-wound package. On the basis of the device used to reassemble the separated fibres, distinctions are drawn between:

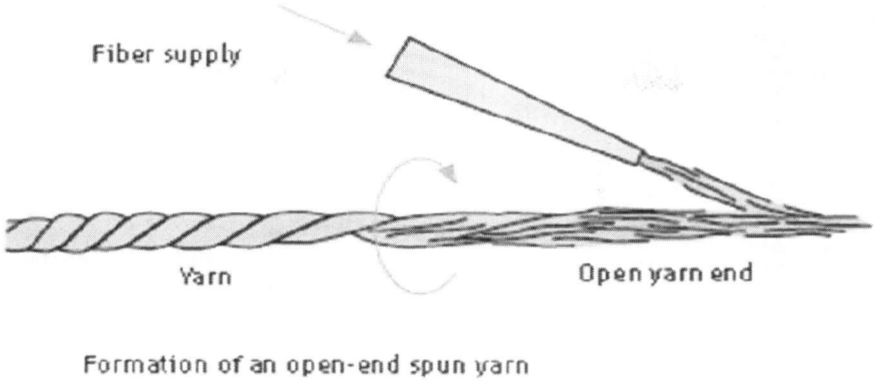

Figure 5.1 The schematic illustration of OE spinning

5.3 Principle of friction spinning

The fundamental difference between open-end friction spinning and open-end rotor spinning is the way in which fibres are collected and twisted onto the tail end of the seed yarn. In friction spinning, the fibres are not collected to form a fibre ribbon that is then twisted. Instead, the fibres are individually collected and twisted onto the yarn. Two rotating, perforated, cylindrical rollers insert the twist by frictional rolling of the yarn tail while simultaneously twisting fibres onto the yarn tail.

Friction spinning uses two friction surfaces to roll up fibres into a yarn. A simplified sketch of friction spinning is shown in figure 5.2.The fibres flow freely to two rotating friction drums (they are also known as spinning drums, friction rollers, torque rollers). The surfaces at the nip of the two drums move in opposite direction to twist the fibres collected in the nip. The yarn is formed from inside outwards, by the superimposition of twisting of individual fibres. The yarn is then withdrawn from the nip to take-up package.

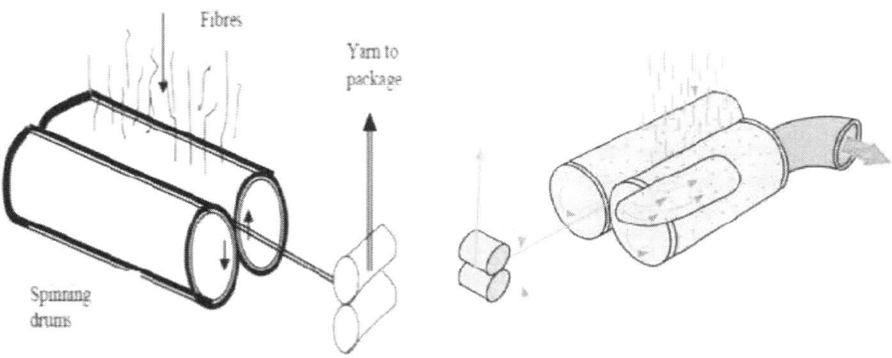

Figure 5.2 Schematic representation of Principle of friction spinning

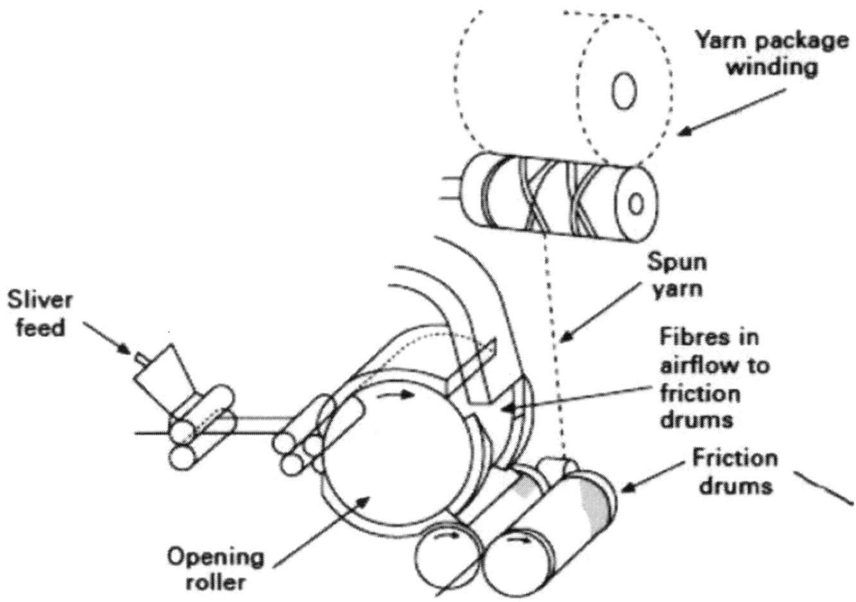

Figure 5.3 The detailed schematic diagram of friction spinning unit

Friction spinning is an alternative open-end spinning method in which friction, rather than a rotor, is used to insert twist. An opening roller attenuates a feed sliver, which is then fed into the groove formed by two perforated drums rotating in opposite directions (Figure 5.3). The fibres are held against the surface of the drums by suction. The tail of the forming yarn is also held in the groove by suction, and as the individual fibres pass through the drums, they are bound to the tail as a result of frictional forces generated. Twist is inserted because of the friction with the two drum surfaces, and the completed yarn is continuously drawn away in the direction parallel to the axis of the friction drums and delivered to a package-forming unit.

This process is included in the open-end group because the fibre strand (draw frame sliver) must be opened completely into individual fibres and then reassembled to a new strand (yarn). The formation of a new strand is carried out by using suction to bring the individual fibres into engagement with the rotating open end of the yarn, e.g. by perforated drums with an internal vacuum. Binding-in fibres and imparting strength are effected by continuous rotation of the yarn end in the converging region of two drums.

The rotation of the yarn end arises from the rotary movement of the two drums and is generated by frictional contact at the drum surface. The yarn formed in the convergent region by collecting fibres and binding them in can be continuously withdrawn and wound onto a crosswound package.

The fineness of the resulting yarn is determined by the mass of fibre feed per unit of time and the withdrawal speed of the yarn; the number of turns is determined by the relationship between yarn end revolutions and withdrawal speed.

The rate at which twist is imparted to the yarn is markedly lower than that which would be expected from the rolling of the yarn end between the two drums. This fact, often attributed to slip, is the result of the very complex details of the yarn formation process.

The economic and technological limits of friction spinning and rotor spinning systems are in approximately the same count range. They are direct competitors in the marketplace.

5.4 Classification

The operations to be carried out in this spinning process are the same as those required for rotor spinning:

1. opening of the fibre strand;
2. acceleration of the fibres;
3. collecting the fibres into a new strand;
4. imparting strength by twisting;
5. withdrawal of the resulting yarn;
6. winding onto a cross-wound package.

Opening is performed by the elements that are same as the ones already used for this purpose in rotor spinning. Collection of fibres can be performed on moving or stationary surfaces, and twisting can be effected by a transfer of forces from some kind of surface. Several different kinds of collection procedure and many different types of surface can be used.

Accordingly, there is not just one kind of friction-spinning system, but there are several. The fibres come into contact with a surface that is moving more slowly than they are.

The most widely used types are those with the following characteristics:

1. single-sliver feed;
2. one opening roller;
3. friction assembly also acting as collection device;
4. two friction surfaces;
5. two perforated drums or one perforated drum and one
6. blind drum in combination.

5.5 Technological relationships

Feed

Multiple-sliver feed improves evenness but also leads to high costs and the need for a very high degree of opening.

Opening

Opening is performed as for rotor spinning. In this case also, straightening of the released fibres and the degree of longitudinal orientation are problematic, but these two aspects exert a strong influence on yarn characteristics.

Fibre transport

The fibres can move to the collecting device in free flight (airborne) with (Platt Saco Lowell Master spinner) or without (Dref-2000) guidance by a duct. Free flight of the fibres without guidance in a duct leads to fibre disorientation, which affects not only the yarn characteristics but also the spinning limits.

Fibre collection

The fibres are drawn by a suction airstream toward the collecting surface and the open yarn end (Fig. 5.4 (a), (b), and (c)). In rotor spinning, the fibres are additionally accelerated during collection and are thereby straightened, but in friction spinning the opposite happens.

In terms of flow direction, the fibres meet the drums and the open yarn end at right angles to the direction of yarn withdrawal (Dref), in the same direction, or in the opposite direction.

In accordance with the system described by Luenenschloss and Brockmanns, reference is made to forward (Fig. 5 (b)) or backward (Fig. 5 (c)) spinning. In general, fibre guidance can perhaps be classified into (refer to Fig. 5.4):

1. right-angle guidance (a);
2. forward guidance (b); and
3. backward guidance (c).

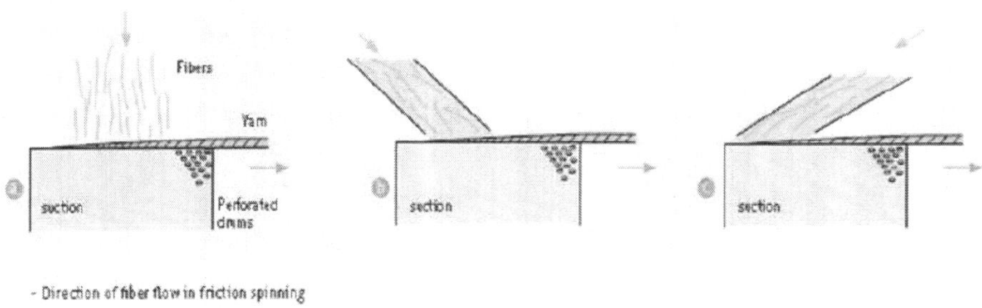

Figure 5.4 Schematic illustration of directional flow of fibres

The result is fibre-buckling and deterioration in fibre orientation. The fibres are bound into the yarn in a loop form [1]; this effect is clearly visible in the yarn product and is more marked with longer fibres. The strength of friction-spun yarn is therefore lower than that of rotor-spun yarns.

Imparting twist

The transfer of rotation to the yarn is dependent on the coefficient of friction and the contact pressure; both these quantities are difficult to keep constant between spinning positions and over time. The apparent slip is variable. A notable characteristic of friction-spun yarn is therefore uncertainty about the rate of imparting twist. Nevertheless, from the technical and economic points of view, this method of imparting twist exhibits remarkable advantages. In practically all other twisting assemblies, one revolution of the twisting element is needed to impart one turn of twist to the yarn. In friction spinning, one revolution of the twisting element can generate several turns of twist. This result is obtained because of the large difference in diameter between the drums and the yarn.

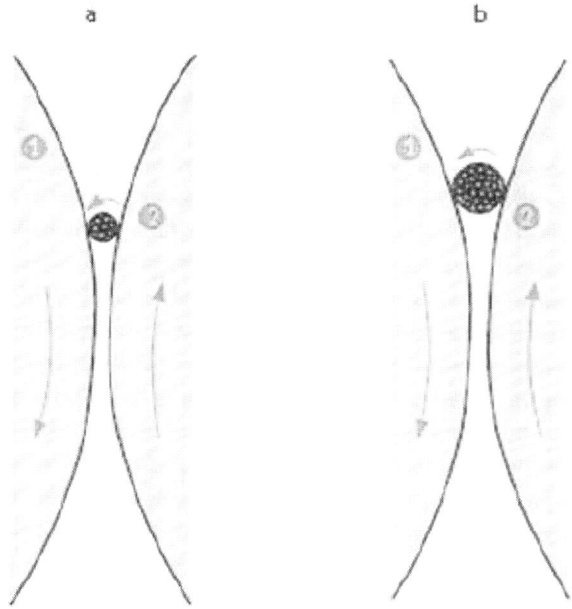

Fine and coarse yarns in the convergent region of friction-spinning drums

Figure 5.5 Schematic illustration of the twist insertion zone

The high transmission ratio (up to 200:1) has the further advantage that a lower rate of drum revolutions suffices, although, when considered in relation to the diameter ratio, the yarn takes up only 15 - 40 % of drum rotation.

With reference to Fig. 5.5 (a) and (b), drum (1) has to rotate through a fraction of a revolution to cause the yarn to rotate once, i.e. one full drum revolution generates 100 and more yarn turns. The illustration also shows that the transmission ratio is still greater for fine yarns (with a smaller yarn diameter) than for coarse yarns.

In the course of one drum revolution, the fine yarn therefore takes up more turns of twist than the coarse yarn. This remains true even though the smaller zone of contact of the finer yarn on the drums leads to greater apparent slip. This is the only spinning method in which the delivery speed is practically independent of yarn count.

Withdrawal and winding up

In contrast to most other spinning processes, yarn tension (and hence end break frequency) is very low during withdrawal from the spinning zone. Tension therefore has no influence on the spinning limit.

The yarn is wound up onto cross-wound packages so that, in comparison with conventional spinning, rewinding is eliminated. Like any other one or more carded slivers are passed to the main opening roller (2) (i.e., a drum clothed with saw teeth) after leaving a drafting arrangement (1). While the drafting arrangement has only a slight drafting effect, the saw tooth roller opens the strand into individual fibres. The fibres separated in this way are lifted off the roller by a blower (3) and form a cloud (7), descending toward two perforated drums (4). One suction stream (5) per drum draws the fibres into the convergent region between the drums. The open end of the yarn (6) projects into this zone and is also sucked toward the perforated drums.

Advantages are as follows:

1. high delivery speeds;
2. low yarn production costs (lower than those of ring spinning);
3. elimination of rewinding;
4. low end breakage rates;
5. yarn character similar to that of ring-spun yarn;
6. no wrapping fibres;
7. optically good mass evenness (well suited to knitted goods);
8. better and softer handle than that of rotor-spun yarn;
9. smooth yarn appearance.

Disadvantages are:

1. low yarn strength;
2. high tendency to snarl;
3. higher number of fibres needed in yarn cross-section;
4. difficulty of keeping spinning conditions constant;
5. high air consumption;
6. increasing unevenness and imperfections with increasing spinning speed, and further reduction in yarn strength.

Delivery speeds can be made correspondingly high. Spinning speeds of 500 m/min or even higher are conceivable. Unfortunately, the spinning speed is limited in practice by yarn quality to some 200 m/min. In fact, a higher fibre throughput rate leads to a deterioration in yarn quality.

The friction-spinning system is based on the principle of open-end spinning, which consists of the following operations: feeding a sliver of fibres into the spinning system, separating the sliver into individual fibres (i.e. opening), reassembling the individualized fibres, twisting the reassembled fibres to form the yarn, and winding the yarn onto a bobbin to form a yarn package.

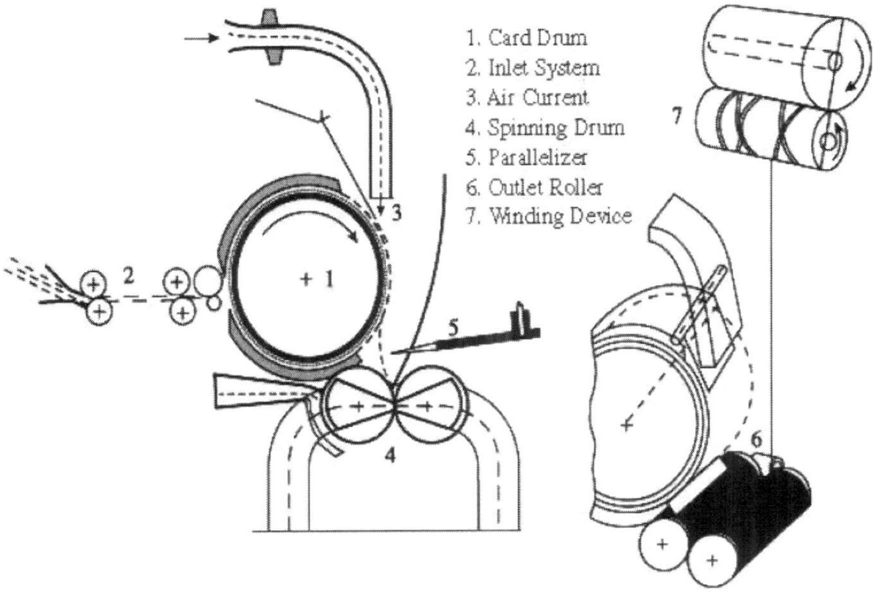

Figure 5.6 DREF spinning system

In an open-end friction-spinning process, a feeding channel is provided to convey the individualized fibres from an opening cylinder into the nip of the yarn forming zone. The fibres are twisted together into a yarn, alongside a suction slit in the yarn forming zone generally formed by two friction rollers, in close proximity with each other and driven in the same direction. The process is assisted by air suction through the roller perforations. The resultant yarn is withdrawn from the nip of the friction rollers and wound on a package. In the design of the Master-spinner the sliver is opened into individual fibres by a pinned beater and then transferred by a transport channel to the yarn forming zone at a 25–28° angle by means of suction through the perforated roller. Two friction rollers are used to insert twist into the reassembling fibres: one is a hollow roller having a perforated surface and an internal suction slot positioned parallel to the nip line, and the other is a solid roller, which provides effective friction transfer. The Master-spinner is capable of processing fibres up to 40 mm long, producing yarns over the count range of 15–59 Tex at delivery speeds of up to 300 m/min. The DREF-2 friction spinning machine essentially consists of a specially designed feeding system, which retains the slivers and provides the required draft. These drafted slivers are opened into individual fibres by a rotating opening roller. The individualized fibres are transferred into the nip of two perforated friction rollers, where they are held by suction. The fibres are subsequently twisted by mechanical friction on the surfaces of the friction rollers. The low yarn strength and the requirement of having more fibres in the yarn cross-section have restricted DREF-2 spinning to the coarser count range of 98–1181 Tex. Unlike the DREF-2 and the Master-spinner, the DREF-3 is a core–sheath type of friction-spinning arrangement. On this machine, there are two drafting units, one for the core fibres and the other for the sheath fibres.

This system produces a variety of core–sheath types of structure and multicomponent yarns, through selective combination and placement of different materials in the core and sheath in the count range of 33–591 Tex, with delivery speeds as high as 300 m/min. With the DREF-5 friction-spinning machine, the individualized fibres from a single sliver are fed through a fibre transport channel into the yarn forming zone at an inclined angle to the yarn axis. At present, the DREF-5 system is the optimum technical solution for the production of the best quality friction yarns within the count range of 15–37 Tex, with production speeds of up to 200 m/min.

The DREF-2000 friction-spinning machine employs the classic DREF-2 system. This machine can produce S- and Z-twisted yarn without mechanical alterations to the machine by the operator. Yarns of 41 Tex can be produced at production speeds of 250 m/min. DREF-2000 friction spinning machines are utilized for recycling textile wastes as well as the spinning of technical and other yarns. The DREF-3000 friction spinning machine, the latest model of the machine, is utilized especially for the production of multi-component yarns (hybrid yams) of count 33–666 Tex with production speeds of up to 250 m/min. This machine has a filament guide and is capable of operating with several yarn cores, which are axially fed to the spinning drums. These core yarns are then provided with a fibre sheath and are positioned precisely in the middle of the yarn.

A high twisting speed can be obtained even while using a relatively low speed for the friction drums, because the friction drum diameter is much larger than that of the yarn. Because the yarn is withdrawn from the side of the machine, fibres fed from the machine end away from the yarn delivery point tend to make the yarn core, while fibres fed from the machine end closer to the yarn delivery point tend to make the sheath. This characteristic can be conveniently used to produce core-sheath yarn structures for a specific purpose, such as a yarn with the strength of a polyester core and the natural feel of a cotton sheath. Additional core components, filaments or drafted staple fibres, can also be fed from the side of the machine, while fibres fed from the top of the machine, the normal input, form the sheath. The fibre configuration in friction spun yarns is quite different from that of other yarns. When the fibres come to the friction drum surfaces, they decelerate sharply from a high velocity to become almost stationary. This causes fibre blending and disorientation. Due to the very low tension in the yarn formation zone, fibre binding within the yarn is also poor. Slub yarns, which are potentially important for decorative effects, can be manufactured on the friction system by changing the feed rate of one or more of the slivers, or by injecting the fibres directly into the friction zone. However, the yarn tends to offer poor performance in processing and use, as a result of the poor binding of the fibres in the yarn indicated earlier. Friction spinning is not often used for cotton spinning because the structure of the yarn (in terms of fibre orientation, fibre packing, etc.) results in lower strength. It is used for producing coarse yarns for industrial applications, and particularly technical textiles.

Because of the very large ratio between the drum and yarn diameters, the rotational speed of the drums need not be high, provided adequate twist efficiency is achieved. The twist efficiency is reduced due to the slippage between the yarn in the nip and the drum surfaces. It is possible to have a twist efficiency as low as 40%. But even allowing for this, friction spinning is still the most efficient way of inserting twist to fibres, because twist is directly applied to yarn end.

Unlike ring spinning and rotor spinning, friction spinning imposes very little tension to the yarn. So the ends-down rate in friction spinning is very low and the yarn can be withdrawn from the nip of the drums at a very high speed, say 250 m/min. This makes friction spinning more productive than ring and rotor spinning.

Similar to rotor spinning, friction spinning uses sliver feed and tooth drafting. Fibres opened by a toothed roller are directed towards the nip of the friction drums, at a very high speed. The fibres should then impinge on the friction surface that is entering the nip or the rotating mass of fibres in the nip. Because the velocity of the entering fibres is much higher than the surface velocity of the drum surface and the rotating mass of fibres in the nip, fibres are decelerated as they impinge on the drum surface or the rotating fibres in the nip. This deceleration causes considerable fibre buckling just before the fibres are incorporated into the yarn structure. As a result, the fibre alignment in friction spun yarns is poor, leading to poor strength of friction spun yarns. Having long fibres does not help yarn strength much in friction spinning, because the long fibres buckle more readily than short ones, so their configurations within the yarn structure may not be as good as shorter fibres. The poor yarn strength also means that friction spinning can only produce relatively coarse yarns. With friction spinning, a core component can be easily introduced in the nip to make a composite yarn of a sheath/core composition.

Examples of this will be discussed in the following section on DREF friction spinning systems. In this section, we have discussed the basic principle of friction spinning, and the key features of friction spun yarns. Next, we will discuss the DREF 2 and DREF 3 friction spinning systems developed by the Fehrer Company located in Linz, Austria.

5.6 DREF 2 friction spinning system

The DREF 2 friction spinning system was introduced into the world market in 1977. It is designed for coarse yarn counts in the 100 Tex to 4,000 Tex range. The DREF 2 system is primarily used for the recycling of all types of textile waste fibres and mixtures with 10 -120 mm fibre lengths, and the spinning of technical and other yarns for specialised applications, such as blankets, cleaning rags and mops, yarns for secondary carpet backings etc. A diagram of the DREF 2 friction spinning system is shown in figure 5.7. As mentioned in the previous section, toothed drafting is used in friction spinning. With the DREF 2 system, the feed slivers are opened and drafted by the teeth of a carding drum. The individualised fibres are then stripped from the carding drum by centrifugal force, supported by an air flow. Gravity and air flow then carry the fibres into the nip of two perforated spinning drums. Assisted by air suction through the spinning drums, the fibres in the nip are twisted by friction on the two drum surfaces to form the yarn.

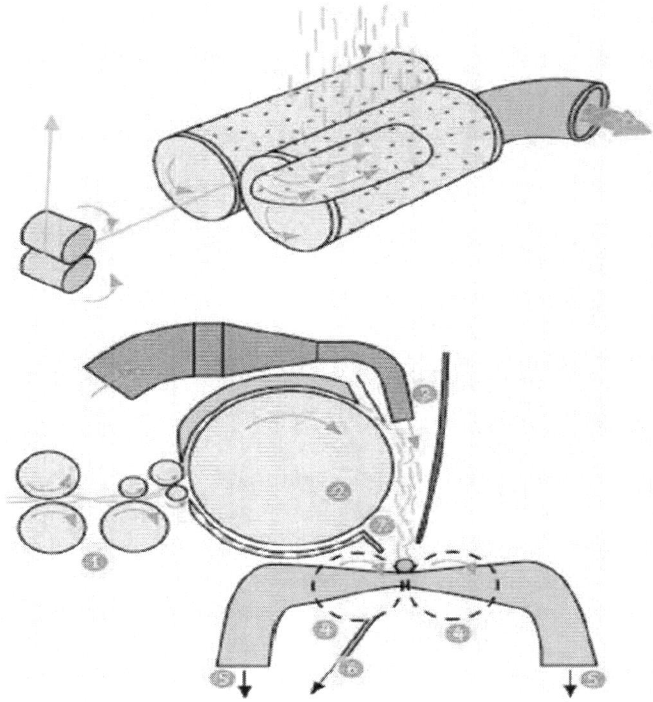

The Dref-2000 spinning system

Since these rotate, the yarn also rotates in the convergent region. The newly arriving fibres contact the rotating yarn and are thereby caught and twisted in. It is only necessary to withdraw the yarn continuously to twist fibres newly arriving in the convergent region into a yarn. Dref- 2000 is primarily suited to the production of coarse yarns (of medium to long staple fibres) and recycling yarns.

The configuration of the Master-spinner is shown in Fig.5.8. A draw frame **sliver (2)** as normally produced in short-staple spinning mills runs from a **can (1)** into an opening assembly. This consists of a **feed roller (3)** and **an opening roller (4),** and opens the fibre strand in the same way as the opening device in rotor spinning. The separated fibres pass through a specially shaped **fibre channel (5),** carried by an air flow from a vacuum inside **the suction roller (6)** into the converging region between the two friction rollers. As previously mentioned, one of these rollers is perforated to act as a suction roller (6), whereas the second roller is solid. **A yarn (8)** is formed in the convergent zone by the method already described and passes via **delivery rollers (9)** and **winding rollers (10)** to a **cross-wound package (11).**

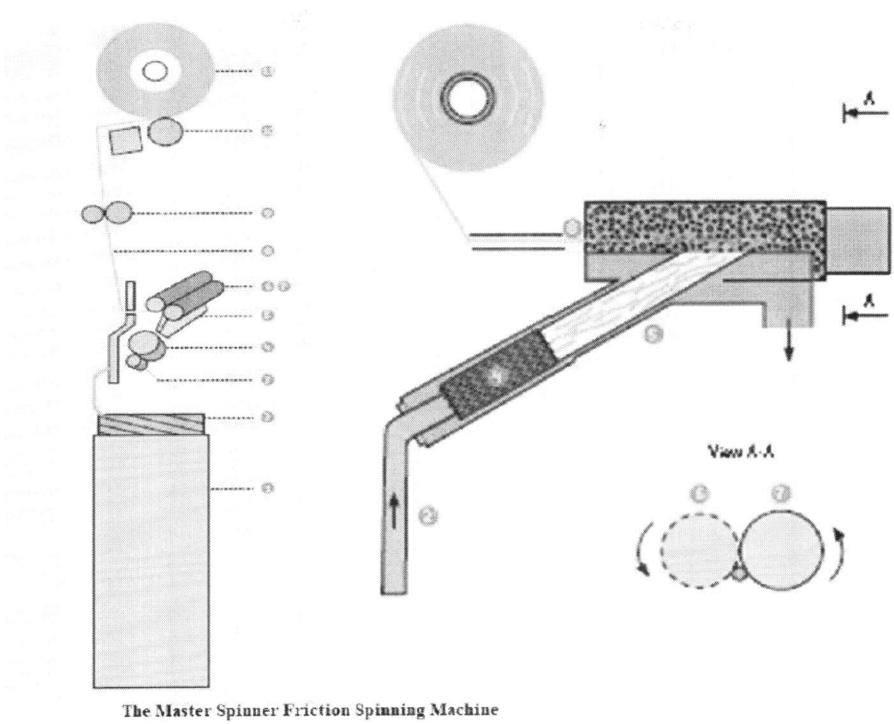

Figure 5.8 The schematic configuration of master spinner

In order to reduce slippage and improve the twist efficiency, the concept of enclosing the fibre assembly between two perforated friction rollers was devised, and formed the basis for the commercial development of two designs of DREF spinning machines, known as DREF-2 and DREF-3.

The DREF-2 machine was demonstrated at the 1975 ITMA exhibition and then commercialized in 1977. The DREF-3 machine was introduced on the market at the end of 1981. The DREF-3 machine was developed basically to improve the yarn quality, extend the yarn count towards the finer end of the count range (up to 33 Tex), and produce multicomponent yarns.

The Dref-2 system can be used to advantage in the spinning of core spun yarns. The deposition and twist of fibres onto the yarn tail provide the opportunity for the yarn tail to be replaced by a filament core, which would then become fully covered by a staple sheath as fibres are deposited and twisted onto the filament.

In this situation, the continuous filament yarn would pass from a filament package, through a thread tensioning guide, along the V-shaped grooved formed by the spinning drums, and via delivery rollers to the package build device.

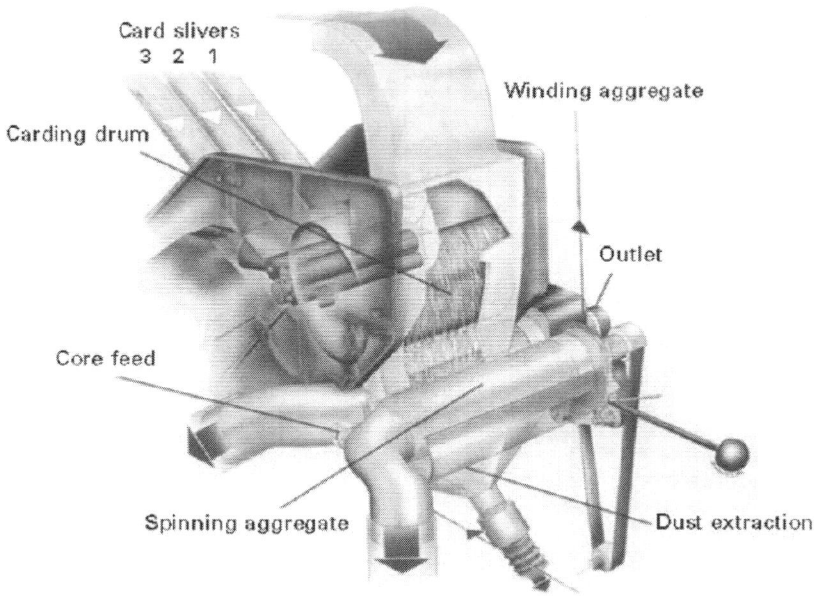

The DREF-2 friction-spinning machine

Friction spinning is defined as an 'open-end' spinning method (or an OE spinning method), in which the yarn formation takes place in the yarn forming zone consisting of two friction rollers with the aid of frictional forces. In traditional spinning systems the fibre supply is first formed into a strand of fibres, almost equal to the final count of the yarn, and then twist is inserted to form the yarn.

With OE friction spinning the fibre supply must be opened completely into individual fibres and reassembled at the nip line of the two friction rollers to form a twisted strand of fibres, i.e. the yarn, each fibre being twisted onto the forming strand by the friction rollers as soon as it reaches the nip line.

While the production speeds are limited with both ring and rotor spinning, the friction spinning system has a low spinning tension and is therefore suitable for the production of yarn at high speeds of up to 500 metres per minute.

The major disadvantages are inadequate yarn strength, i.e. low utilization of the fibre properties; and inconsistency of the spinning results. The yarn is then withdrawn by the takeup rollers at delivery speeds of up to 250 m/min, and wound onto a large yarn package. A filament core can be easily introduced into the nip of the spinning drums via the core feeding, to make a composite yarn of a sheath/core structure. During spinning, the filament core gets false twisted by the spinning drums, while the staple fibres are deposited on the false twisted filament to make a sheath. The staple fibres are twisted as usual. But as the filament core emerges from the nip of the spinning drums, the false twist in it is removed automatically, and the sheath fibres receive a reserve twist in the process. The resultant composite yarn has the characteristics of a twistless filament core surrounded by a sheath of helical wound fibres of varying helix angles. The core/sheath effect can also be achieved without the filament component. As indicated in the previous section, the yarn in the nip of the spinning drums has a tapered end, and fibres deposited in the thin end of the taper are likely to end up in the core position of the resultant yarn. For example, fibres in the leftmost card sliver are likely to stay as core fibres in the yarn, surrounded by sheath fibres from the remaining two card slivers. This preferential fibre arrangement facilitates an economic use of a variety of raw materials. A 'core' sliver of waste fibres may be used with other 'sheath' slivers of high quality virgin fibres to make a quality yarn with reduced raw material cost. It is worth mentioning that even if a filament core is not to be used as part of the final yarn, a filament is often used to help start the spinning process. Once started, the filament is then cut to allow the process to continue without it. This also applies to the DREF 3 friction spinning system that is discussed next.

The DREF-2 friction spinning machine essentially consists of a specially designed feeding system, which retains the slivers and provides the required draft. These drafted slivers are opened into individual fibres by a rotating opening roller. The individualized fibres are transferred into the nip of two perforated friction rollers, where they are held by suction.

The fibres are subsequently twisted by mechanical friction on the surfaces of the friction rollers. The low yarn strength and the requirement of having more fibres in the yarn cross-section have restricted DREF-2 spinning to the coarser count range of 98–1181 Tex.

5.7 DREF 3 friction spinning system

After the introduction of DREF 2 into the world market in 1977, Dr Ernst Fehrer began work on the development of the DREF 3 friction spinning system, which was first presented to the public at the 1979 international textile machinery exhibition (ITMA'79) in Hanover. In 1981, DREF 3 entered the global textile machinery market. DREF 3 is designed for the manufacture of multi-component yarns in the medium count range (25 - 667 Tex). The yarns have been used in a wide range of industrial applications, including fire-resistant protective clothing, aircraft and contract carpeting, conveyor and transport belts, composites for aviation and automotive industries etc. Figure 3.4 shows a diagram of the DREF 3 friction spinning system.

Unlike the DREF-2 and the Master-spinner, the DREF-3 is a core–sheath type of frictionspinning arrangement as shown in Fig. On this machine, there are **two drafting units,** one for the core fibres and the other for the sheath fibres. This system produces a variety of core– sheath types of structure and multi-component yarns, through selective combination and placement of different materials in the core and sheath in the count range of 33–591 tax, with delivery speeds as high as 300 m/min.

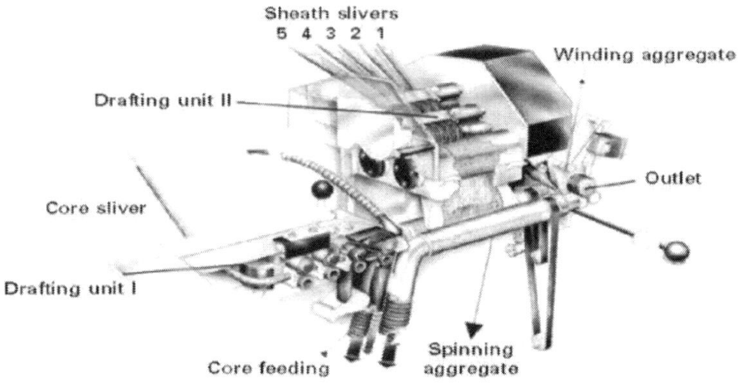

The DREF-3 friction-spinning machine

Figure DREF 3 friction spinning system (Fehrer AG)

There are two modes of fibre feed, namely vertical feed and inclined fibre feed. DREF-2 and DREF-3 have vertical feed systems, whereby the fibres are fed at right angles to the yarn axis. The Master spinner employs an inclined fibre feed known as the backward-feed system, but the DREF-5 unit has an inclined fibre feed known as the forward-feed system.

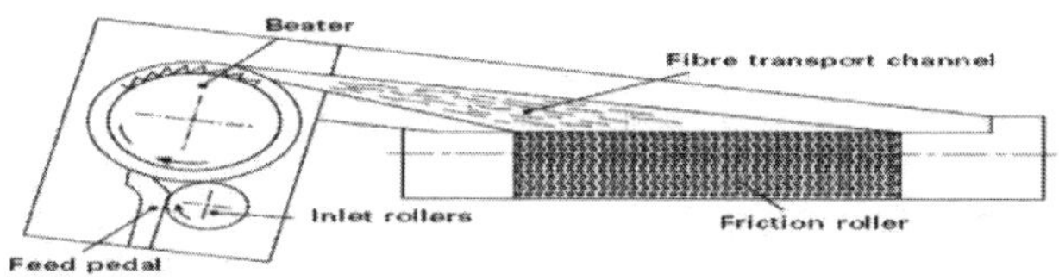

Schematic diagram of the DREF-5 friction-spinning machine

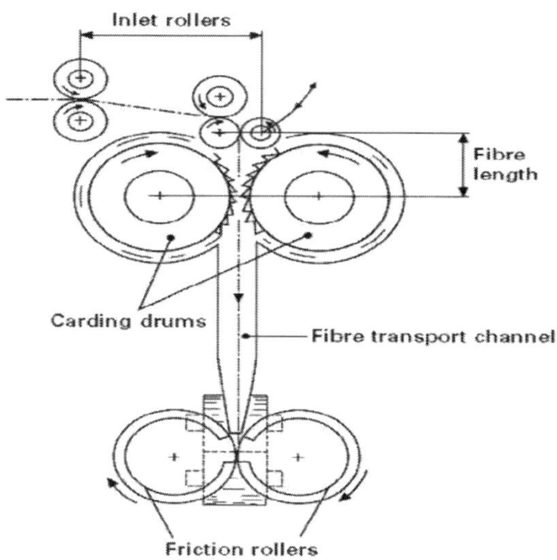

The fibre-opening and feeding device of the DREF-3 friction spinner

The mode of fibre feeding has a *definite effect on the fibre extent and fibre configuration in the spun yarn* and ultimately on the yarn properties. The inclined fibre feed offers advantages such as better fibre-length utilization and the spinning of finer yarns. However, the vertical feed results in the production of stronger but coarser yarns.

There are a number differences between DREF 2 and DREF 3. First of all, toothed drafting is achieved by two toothed drums in DREF 3 rather than just one carding drum in DREF 2.

Second, another roller drafting unit is now added in DREF 3. This drafting unit will deliver parallel fibres that will form a core of parallel fibres in the final yarn, surrounded by the sheath fibres from drafting unit II. There is also the option for the introduction of a filament core as in DREF 2. Therefore, composite yarns of three different components can be engineered on the DREF 3 system. Figure 5.7 shows a side view of the DREF 3 system.

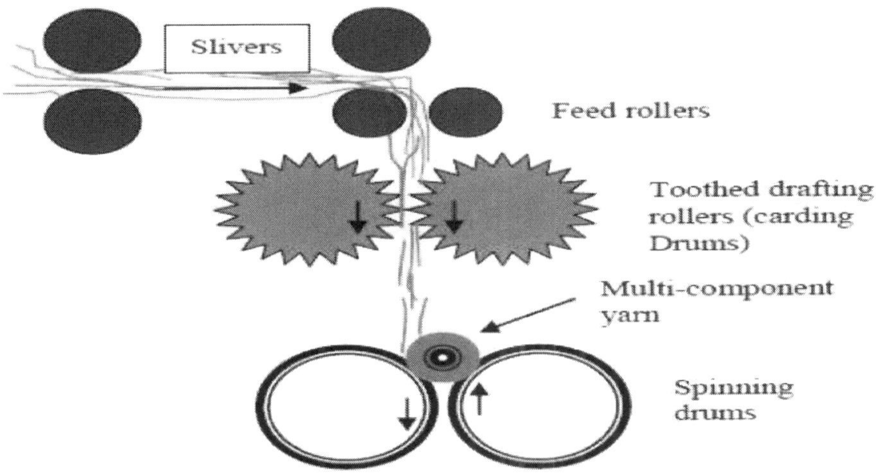

Figure 5.9 A side view showing the principle of DREF 3

For good doubling effect, fibres slivers are used as feed slivers. The overall density of the feed slivers is very high, which means they have to be fed to the toothed drafting rollers (carding drums) at a very slow speed. To minimise fibre damage, the distance between the clamping line of the last pair of feed rollers and the line of the narrowest clearance between the two toothed drafting rollers is set at about one fibre length. In addition, this distance is adjustable to cater for raw materials of different fibre length.

As with DREF 2, fibre buckling occurs as the individual fibres impinge upon the surface of the spinning drum and the mass of fibres already in the nip of the two spinning drums. This leads to poor fibre orientation in the yarn, which reduces yarn strength. But in DREF 3, the core fibres are of parallel configuration. So DREF 3 yarns should have higher tenacity than DREF 2 yarns under similar conditions.

The percentage of core/sheath components can be easily adjusted with the DREF friction spinning systems.

Various fields of application of friction spun yarns are broadly categorized as *industrial textiles, transportation and automotive textiles, safety and protective textiles, outdoor textiles, outerwear textiles and domestic textiles.* In industrial textile fields, DREF friction spinning is ideal for the manufacture of non-asbestos yarns for application in heat-, fire- and flameretardant clothing, friction linings (clutch and brake linings) for the automotive industry, gaskets, packing materials, insulation tapes, braided tubing, cables, wires and ropes.

Chapter 6

Air-Jet Spinning

6.1 Introduction

There are many different spinning systems currently being explored in the textile industry. Some of them are in experimental stage, many are still commercial use, and some of them have been withdrawn from the market. Amongst all, the conventional ring spinning method is the most widely used, accounting for an estimated 90% of the world market spinning machines. Ring spinning is providing all fibres to be spun into a wide range of yarn count with the lowest rate of yarn faults with the best quality and hence, this spinning technology is still the most widely used one in the market.

Compact spinning is one of the modifications of ring spinning process which does better integration of the constituent fibres into the spun yarn structure. The fibre bundle gets condensed by air suction, and hence, this resultant yarn has better tensile properties for the same twist level, lower hairiness and better yarn evenness. Rotor spinning is another most commonly accepted unconventional short-staple yarn spinning technology currently by the textile industry. It is an open end spinning process in which the input fibrous material to the spinning system is highly drafted, ideally to the individual fibre state. The individual fibres are subsequently collected onto the tail end of a seed yarn that is rotated to twist the fibres into the yarn structure and thereby form a new length of yarn. The spinning is continuous as the input material is continuously collected onto the open end of a previously spun yarn.

In the early 1980s, air-jet spinning system was launched. Initially only the man-made fibres could be used as the raw material; later, it was improved for cotton yarn spinning as well. Although the developments aimed to produce 100% cotton yarns, the acceptable quality was provided with polyester/cotton blended yarns in terms of yarn strength.

Air-jet spinning systems: Air-jet spinning is the most promising yarn spinning technique at the beginning of the twenty-first century. It is the fastest means of industrial production of staple fibre yarns. The main technological feature that differentiates the morphology of air-jet spinning from the other up-to-date technologies, is the application of swirling airflow for the purpose of inserting a twist into the yarns.

The innovative idea for using air-jet currents as the twisting element for spinning of staple fibre yarns has been tried for decades. In the history of air-jet spinning Goetzfried method, is the first to use air-jet flow as a twisting device. The Goetzfried method was basically an open-end spinning principle: fibres from a sliver were separated and transported to the open end of a spun yarn (as in the case of the later developed rotor spinning method). The control of the spinning and twisting was done by changing the parameters of the airflow, which was a very complicated task from the point of view of the physical process itself.

By 1950s, the swirling airflow successfully used as a twisting element in the process of air jet texturizing (even today this is one of the most popular methods for production of texturized filaments). When the machine was fed with staple fibres (in the place of the continuous filaments), to the nozzle entrance, an unexpected result happened – the fibres were spun into a yarn.

This yarn had a unique and unknown structure: a core of relatively parallel fibres, wrapped by a small quantity of surface fibres (US Patent 1963). The yarn structure was called 'fasciated' yarn (The term 'fasciated' is derived from 'fasce': a bundle of rods, wrapped with ribbons).In 1963 air-jet spinning was commercially through the Rotofil method and machine of E. I. Du Pont de Nemours and Co. which had only one spinning nozzle. However, the yarns were not accepted well in the market.

In 1973 another air-jet spinning technique was demonstrated in Poland and it was termed as PF-1. This spinning process was based on a nozzle configuration of three injectors, which generated a swirl flow in the nozzle chamber. Yarns of 18–63 Tex were delivered with a speed of 80–200 m/min. However, the PF-1 method used the same open-end principle as the Goetzfried method, and was an unstable spinning process and hence, there was no commercial appeal. The same was the result with the PAM-150 method developed in the former Soviet Union. It was not well accepted as a wide industrial application, as the air-jet spun yarns were very weak and needed additional folding and twisting at a delivery speed of only 20–25 m/min.

There was a renaissance of air-jet spinning technology with the development of MJS machine of the Japanese machinery manufacturing company Murata Machinery Ltd (Murata Jet Spinner). This innovative air-jet spinning machine was introduced for the first time at ATMEI 1982 and attracted the attention of experts with its capability of spinning fine count yarns at a production speed of 150 m/min.

The main differences between the three types of machines were: in the number spinning units and the production speed.

In 1985 at ATME-I in Greenville, Murata exhibited the MJS 801 machine, which produced yarns with linear density ranging from 10 to 80 tex, from both cotton and cotton/polyester blends of staple lengths up to 38 The Toray Group presented the AJS 101 (Air-Jet Spinner) machine, which produced air-jet spun yarns with strength about 10% less than the strength of similar ring spun yarns. This result was much better than the reported strength values of MJS yarns, which could reach about 50–60% of the strength of ring spun yarns. Hence, the Toray Group presented the new AJS yarns as suitable for both warp and weft threads, while Murata's MJS yarns could be used only as weft threads and for knitting.

The main constructive difference between the AJS and MJS machines was that AJS used only one spinning nozzle (MJS had two nozzles), which led to a reduction of the energy consumption. Both machines had automatic knotting devices and worked at similar draft ratios (200 times for AJS and 250 times for MJS).

In 1986 the air-jet spinning method had already been assessed as one of the very promising industrial methods for production of staple yarns, particularly medium to fine yarn counts. At the same time 600 MJS machines, producing cotton/polyester yarns for bed linen, were successfully installed and were in operation in the US.

The Spring Industries, US based textile company equipped a whole spinning mill with Murata jet spinning machines and reduced its staff from 450 to 260 persons due to the extremely small number of breaks and easy maintenance.

At ITMA'87 Murata introduced a new version of its air-jet spinning machines: the MJS 802 with the new exclusive nozzle termed as the 'cotton nozzle', which could spin 100% cotton combed slivers into yarns. As a matter of fact this nozzle and the respective machines did not reach wide industrial appreciation/ application because of the high input requirements for the sliver quality. However, Murata dominated in the field in comparison with the machines of Toray (AJS 102, which could not produce cotton spun yarns). Howa and Toyoda did not exhibit any of their former air-jet spinning machines. In their place, the German company Spindelfabrik Suessen participated among the major competitors in the field of air-jet spinning machines.

Murata launched two new air-jet spinning machines namely the 802H and the 804 RJS respectively. The 802H system was equipped with a five line drafting system so as to allow coarse slivers to be processed at high speeds – up to 300 m/min. A modified nozzle, placed closer to the drafting unit (so as to minimize ballooning), was used to assist high speeds The construction of the 804 RJS was similar to that of the 802H except that the second nozzle was replaced with a set of rubber-covered balloon rollers. It was claimed that the new feature would reduce the energy consumption and yarn hairiness, resulting in a structure more similar to the ring spun structure. Although the production speed was very high (up to 400 m/min), the 804 RJS did not become a commercial success.

The development of the air-jet spinning method progressed at a steady pace till 1999, when Murata presented at ITMA'99 its new machine and method – the Murata Vortex Spinner (MVS). Even before their official demonstration MVS machines were installed in the USA. MVS machines were able to produce pure cotton yarns and yarns from cotton and chemical fibre blends at 400 m/min. According to company information, the newly developed nozzle enabled the yarns to be produced with a structure that is very similar to the structure of ring spun yarns.

During the last international exhibition, ITMA 2007, Murata Machinery Ltd (or its textile machinery division Muratec) showed the MVS 61 machine, which was spinning carded cotton slivers into a Ne 40 yarn at a delivery speed of 400 m/min. A new challenge for the Japanese company was the exhibition of vortex units, which were spinning 100% wool staples (19.5 mm) into Nm 48 yarn at 250 m/min. It would be an interesting matter for the future watch if this possibility could be accepted by the worsted yarns market – one of the most conservative in the world of application of new spinning methods.

6.2 Principle of Operation of Air Jet Spinning

Yarn manufacture using the air jet primarily produces fascinated yarns using the false twist principle. Hence, we discuss about the principle of false twisting before going into actual air jet spinning.

False Twisting

Figure 6.1 demonstrates the principle of operation of false twisting. If a fibre strand A is held firmly at two spaced points by clamps K1 and K2and is twisted somewhere between them, this strand always takes up the same number of turns of twist before and after the twist element (T).

However, these turns have opposing directions of twist, which are represented in the example in Figure 1A as Z-twist on the right and S-twist on the left.

If the clamps are replaced by rotating cylinders (Z1 and Z2 in Figure 1B) and the yarn is allowed to pass through the cylinders while twist is being imparted, the result is governed by the false-twist law and is different from the case of the stationary yarns, as previously assumed.

A moving yarn entering the section (b) already has turns of twist imparted in section (a). In the example illustrated (B), there are turns of Z twist.

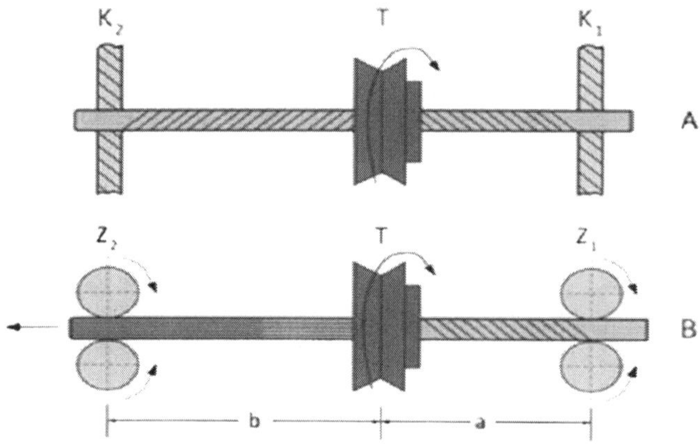

Figure 6.1 Schematic illustration of principle of false twisting

As the twist element is generating turns of S twist in the left hand section, this simply means that each turn of the Z twist imparted in the first section (a) is cancelled by a turn of S twist imparted in the second section (b).

The fibre strand thus never has any twist between the twisting element and the delivery cylinder. In a false-twist assembly, turns of twist are present only between the feed cylinders and the twisting element. This principle is exploited, for instance, in false-twist texturing.

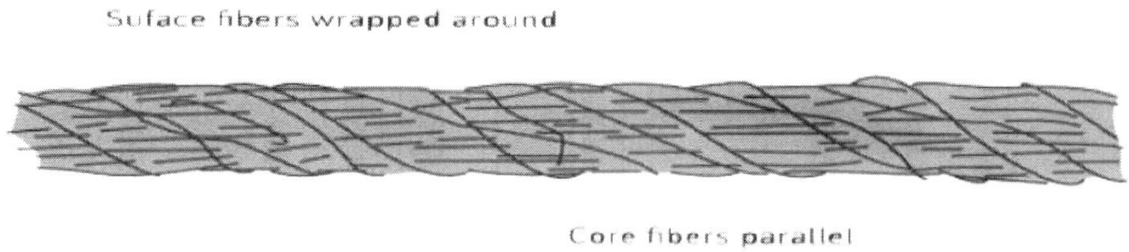

Figure 6.2 Schematic illustration of fasciated yarn structure

Fasciated Yarn through False Twisting:

The idealized structure of the fascinated yarn, as shown in Figure 6.2 consists of parallel fibres held together by wrapper fibres. The wrapper and core fibres are composed of same staple fibre material. Since there is no real twist in the core, this type of yarn structures facilitate high production rates.

The classical air-jet spinning uses the principle of false-twisting to produce a yarn of uniquely different structure from that of ring or rotor spun yarn. While ring-spinning is characterized by a continuity in the fibre flow, and rotor spinning is characterized by a complete separation of fibres prior to spinning, air-jet spinning exhibits an intermediate feature in which part of the fibre strand flows continuously and another part is separated.

Figure 6.3 demonstrates the principle involved in the production of fasciated yarn using the false twisting method. As already explained, the fibres upstream of the false twister have twist which gets cancelled with opposite twist once it passes the false twister leading to no twist downstream of the false twister.

If there are enough edge fibres in the feed fibrous assembly, then these edge fibres do not get twisted with the core fibres upstream of the false twister. Hence, as the core fibres get untwisted after the false twister, these wrappers which had no twist earlier, get wrapped around the core fibres. This produces fascinated yarn structure.

These types of yarn structures were first promoted by DuPont. Figure 6.3 shows the schematic of the DuPont system which did not get commercial success.

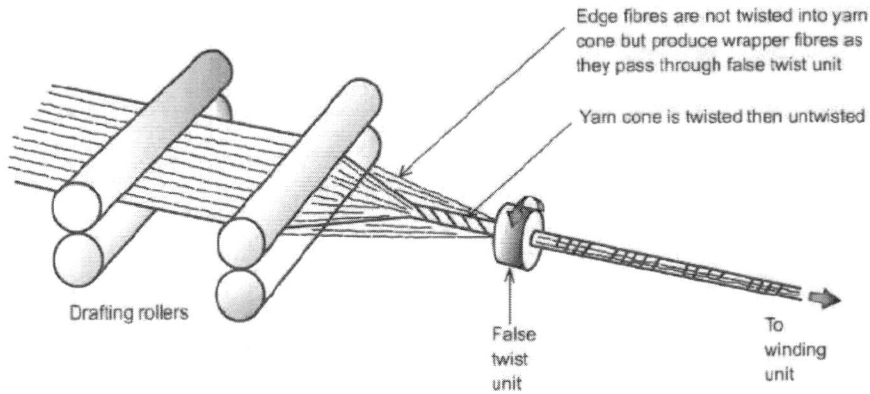

Figure 6.3: DuPont System of Air Jet Spinning

Similar to rotor spinning, the input strand in air-jet spinning is a drawn sliver, which may be carded or combed. Drafting is achieved using multiple zone roller drafting. The consolidation mechanism in air-jet spinning is achieved by blowing out compressed air through air nozzle holes of about 0.4mm diameter to form an air vortex. The air revolves at high speed (more than 3 million rpm). Thus, the rotating element in air-jet spinning is air. This results in a rotation of the fibre bundle at a rate typically ranging from 200,000 to 300,000 rpm.

The figures 6.4 and 6.5 show the air-jet spinning system produced by Murata. Two air nozzles are used: nozzle 1 may be called the "end-opening" nozzle, and nozzle 2 may be called "the twisting nozzle". These names imply the specific functions of these two nozzles as explained below:

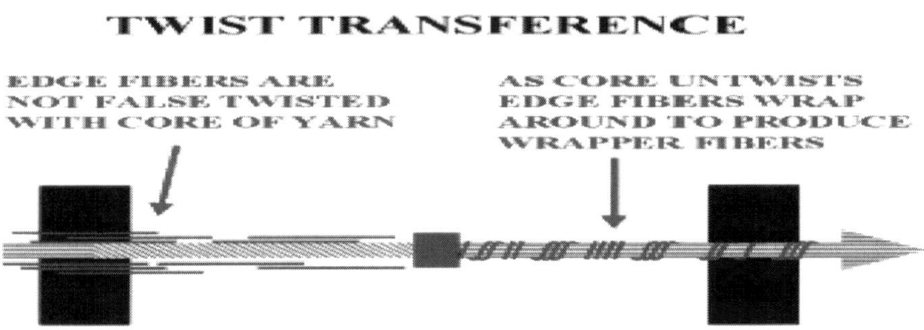

Figure 6.4 Schematic illustration of twist transference in false twisting

To simplify the principle of the consolidation mechanism, suppose that only nozzle 2 is at work and that air is rotating in a clockwise direction. This action will result in twisting the fibres fed to the nozzle to form a yarn. When the yarn leaves the nozzle, untwisting takes place. Thus, with one air nozzle, a case of pure false twisting is achieved. In the actual machine, another nozzle (nozzle 1) is positioned between the nip of the front roller and nozzle 2, with air rotating in a counter clockwise direction. Thus, the two nozzles apply air rotation in two opposite directions. However, the air in nozzle 2 has a higher rotational speed than nozzle 1 to avoid complete false twisting. The fibre strand, coming out of the delivery roll, forms a spinning triangle similar to that in ring spinning. However, fibres in this triangle are under much less tension than those in ring spinning. In other words, the fibres in the triangle are comparatively loose.

The air rotation of the fibre strand in the two nozzles results in ballooning the fibre bundle between the front roller and nozzle 1, and in turning the balloon in nozzle 2. This balloon has no significant tension, which results in some fibres being raised from the bundle surface and move freely. This process is called "the end-opening" action. Thus, the opposite rotation of air in nozzle 1 assists in detaching some fibres from the input strand

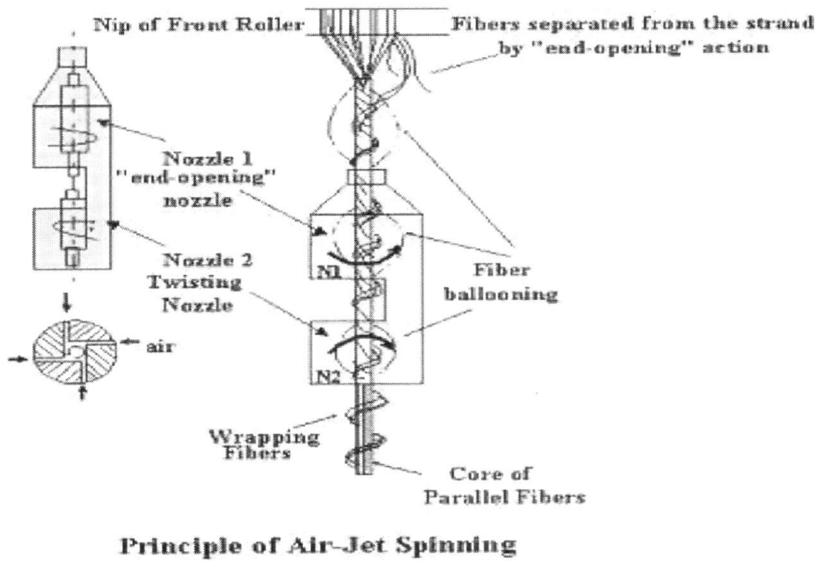

Figure 6.5 Schematic illustration of Murata Air Jet Spinning System

This technique is also known as **fasciated yarn spinning**. There are many variants of the technique. Here, for the purpose of illustration, only the basic technique will be considered. Figure 6.6 portrays a typical air-jet spinning system which consists of a 3-over-3 high-speed roller drafting unit, two compressed-air twisting jets arranged in tandem, a pair of take-up rollers and a yarn package build unit. The basic jet design is also shown. This has a central cylindrical channel (the spinning channel) through which the fibre ribbon from the drafting unit passes.

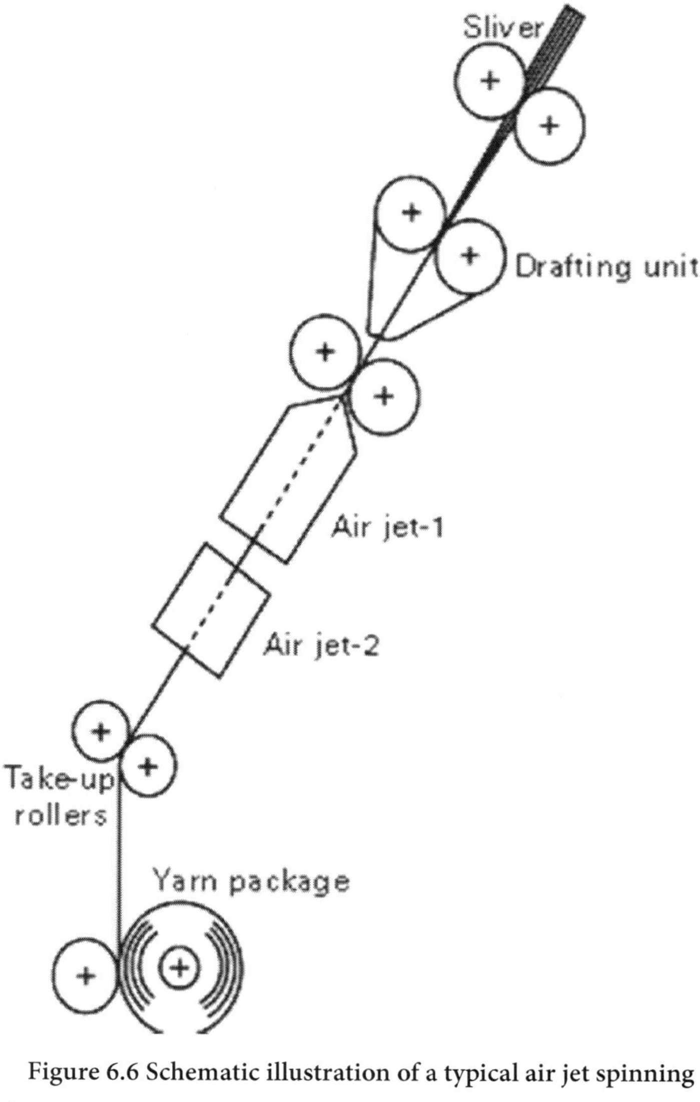

Figure 6.6 Schematic illustration of a typical air jet spinning system

Inclined to the channel axis but tangential to its circumference are four nozzles through which compressed air is injected into the channel, creating a vortex airflow. Each jet of compressed air entering and expanding into the channel has two velocity components of airflow: V1, a circular motion of the air around the channel circumference, and V2, the movement of the air to the channel outlet. The suction at the jet inlet created by *V2 gives* automatic threading-up of the spinning process

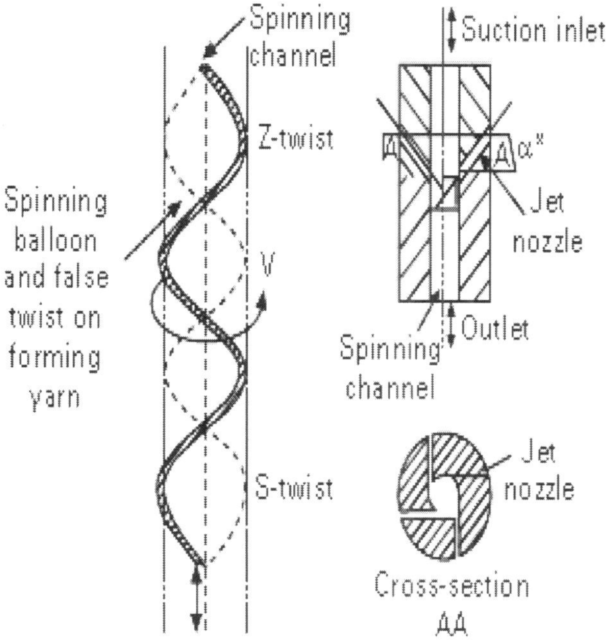

Figure 6.7 Schematic illustration of air jet spinning

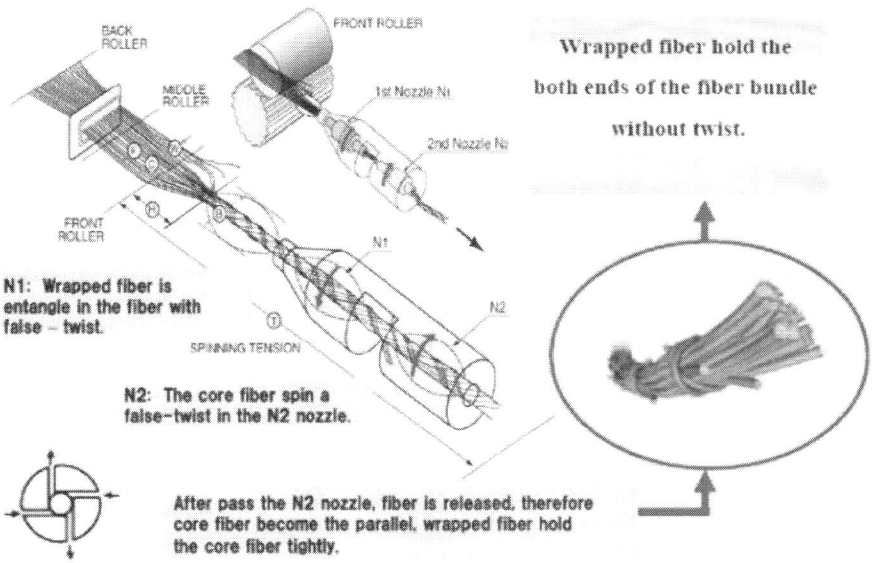

Figure 6.8 Schematic illustration of Air Jet Spinning

The crucial factor when increasing the spinning speed is the twist insertion. Ideally, the mechanically moved mass is reduced to a minimum. Pneumatic methods have hence come into focus in recent years. In addition to higher yarn speed, wear of machine components owing to friction is also expected to be reduced. At present, there are two machine manufacturers offering air-jet spinning machines, Murata (MVS process, introduced 2003, Japan) and Rieter (2011, Switzerland).

6.3 Murata MJS System:

Figure 6.9 shows a schematic of a Murata MJS double nozzle air jet spinning system. The feed material is a draw frame sliver fed from a can (1) which is passed to a drafting arrangement (2), where it is attenuated by a draft in the range of 100 - 200. The fibre strand delivered then proceeds to two air jets (3 and 4) arranged directly after the drafting arrangement. The second jet (4) is the actual false-twist element.

The air vortex generated in this jet, with an angular velocity of more than 2 million rpm, twists the strand as it passes through so that the strand rotates along a screw-thread path in the jet, achieving rotation speeds of about 250 000 rpm. The compressed air reaches the speed of sound when entering the central canal of the false-twist element. Since the axial forces are very low during this rotation, only low tensions arise in the yarn.

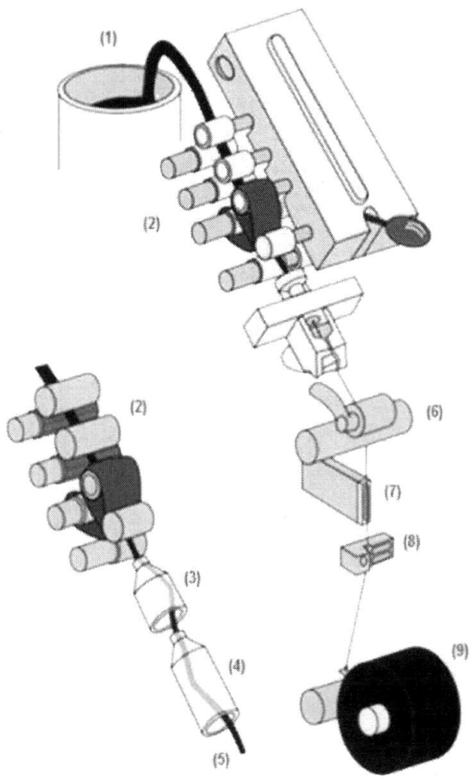

1. Sliver can
2. Drafting arrangement

3 &4. Air-jets

5. Yarn

6. Take-off rollers

7. Yarn suction device

8. Electronic yarn clearer

9. Package winding (CW)

Figure 6.9 : Two nozzle air-jet spinning principle (Murata MJS)

This is partly ensured by causing the strand to emerge from the nip line in a broadly spread form, but mainly by generating in the first jet (3) a vortex with an opposite direction of rotation to the vortex in the second jet (4). This first vortex is in fact weaker in intensity than the second and cannot really affect the core fibres, but can grasp the edge fibres projecting from the strand at one end.

Since the first vortex acts against the twist direction generated by the second jet, it prevents the edge fibres from being twisted into the core or even twists them in the opposite direction around the core fibres. As the strand runs through the second jet, the following occurs.

The turns of twist generated by the jet (4) are canceled in accordance with the false-twist law. The core fibres, i.e. the vast majority, no longer exhibit any twist; these fibres are arranged in parallel.

On the other hand, the edge fibres (which previously exhibited no twist, relatively little twist, or even twist in the opposite direction) receive twist in the direction imparted by the jet (4), as determined by the law of false twist; they are therefore wound around the parallel fibre strand.

They bind the body of fibres together and ensure coherence.

A twist diagram prepared by Dr. H. Stalder demonstrates this twisting procedure (see Figure 6.10).

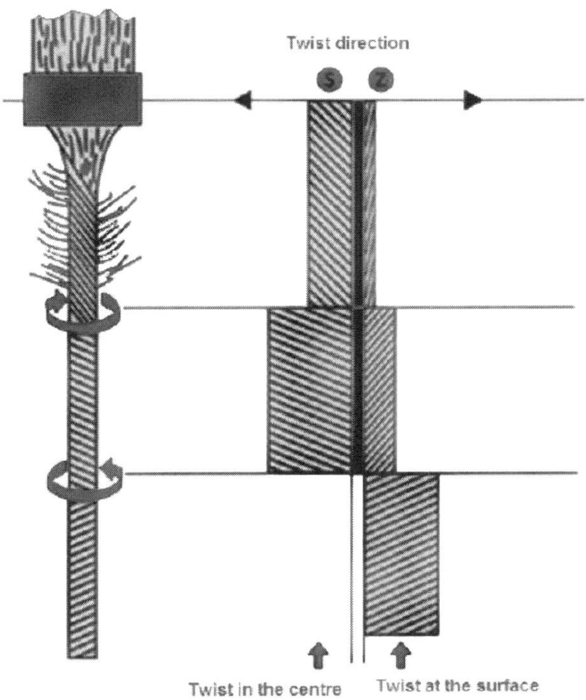

Figure 6.10 Schematic illustration of Twist diagram and twist procedure

The resulting bundled staple-fibre yarn passes from the take-off rollers through a yarn-suction device (7) and an electronic yarn clearer (8) before being wound onto a crosswound package (9). The two nozzle air-jet spinning system represents a very interesting process, which has already been introduced into practical operation with some success.

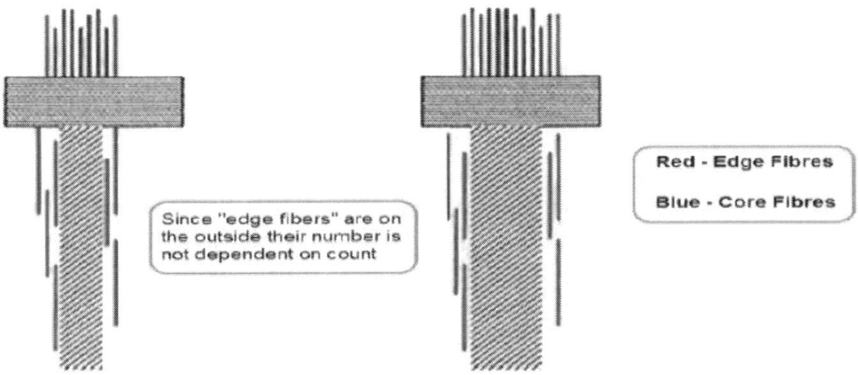

Figure 6.11 Schematic illustration of edge fibres

The tenacity of the fascinated yarns spun with air jet depend on the yarn count. The coarser yarns are weaker than the finer yarns for the same fibre type. Contrary to the expectation, yarns produced with finer fibres show lower tenacity compared to the yarns produced with coarser fibres.

The reason for the above observations is that the strength of the fascinated yarns is derived from the amount of wrapper fibres and the intensity of wrapping. The edge fibres are the ones which ultimately get converted into wrapper fibres. The number of edge fibres are limited to the surface of the yarn and are independent of the number of fibres in the core as shown in Figure 6.11.

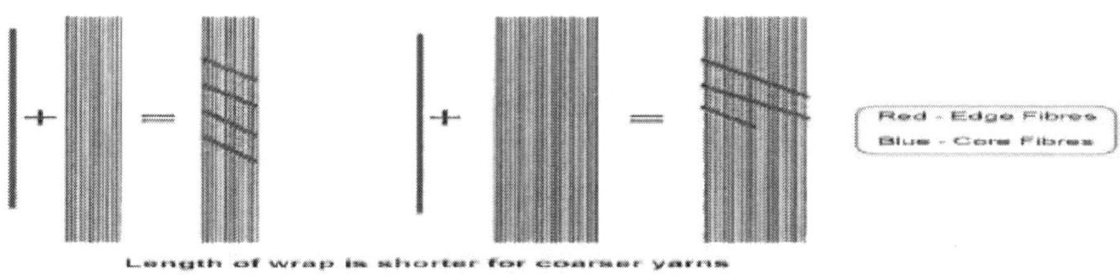

Figure 6.11a Schematic illustration of twisting of coarse and fine yarns

6.4 Air-Jet Spun Yarns :

The air-jet spinning system with distinct way of yarn formation results in the production of unique structure. The air-jet-spun yarn is fasciated yarn consisting of a core of parallel fibres held together by wrapper fibres.

The structure of air-jet-spun yarn is essentially that of comparatively straight central core of fibres held together by taut surface fibres wound onto the central core helically. The straight fibres termed as "core fibres" while the taut, helically fibres called as "wrapper fibres". The fasciated yarn structure was classified into three distinct classes as below

Class 1: In this structure, a part of yarn that has regular helical wrapping and the yarn core is crimped the crimpiness is due to the buckling force generated by wrapping fibre torque and tension. The angle of wrap varies between 40 to 45 degree.

Class 2: This structure has twist less core randomly wrapped by fibres, in singular state and group with angle of wrap varies between 45 to 90 degree.

Class 3: This structure consists of unwrapped section of yarn core, at time having residual twist. The yarn structure resembles that of ring yarn with very low twist in fibres.

There are two methods of air jet spinning: Murata jet spinning (MJS) and Murata vortex spinning (MVS). In MJS, two air jet nozzles are used to twist and entangle the fibres in the sliver. The air vortexes inside the nozzles are in opposite directions. Therefore, the first nozzle twists the fibres in one direction and the second nozzle twists the fibres in the other direction (Figure 6.12).

However, there is no positively controlled twist given to the yarn. Murata vortex spin ning, which is relatively new, was developed for 100% cotton yarns. Yarn does have real twist in its structure, which is similar to twist in ring spun yarns.

Figure 6.12 Schematic illustration of Air jet spinning

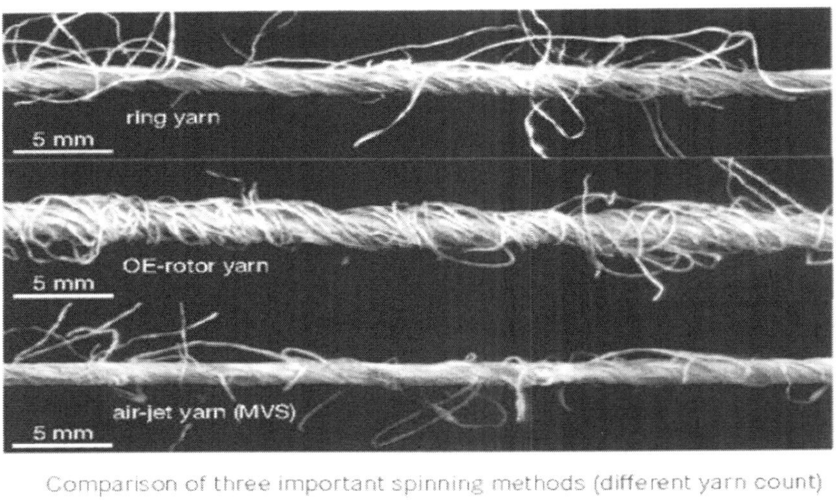

Figure 6.13 Photo-micrographs of the three different types of yarns

6.5 Developments in Air-Jet Spinning

Air-jet spinning machinery may be divided into two main types: single end spinners, and twin spinners In the twin-spinner, two slivers are fed to the same drafting system where they are drafted. The drafted strands are then fed to two different spinning units (air nozzles) to produce two single yarns. These two yarns are then doubled together onto a take-up package suitable for two-for-one twisting system. The twin spinner is therefore suitable for applications where plied yarns are required. In comparison with ring spinning, the twin spinner eliminates roving, winding, and doubling machinery.

Air Vortex Spinning

Vortex spinning technology was introduced by Murata Machinery Ltd. Japan in 1997. This technology is best explained as a development of air-jet spinning, making use of air jets for yarn twisting. The main features of Murata vortex spinning (MVS) are Ability to produce yarn at 400 m/min. which is almost 20 times greater than ring spinning frame production.

1. Low maintenance costs, a fully automated piecing system and elimination of roving frame.
2. The yarn and the fabric properties of MVS yarn are claimed by the manufacturer to be comparable to those of ring spun yarn.

VORTEX yarn is made by the "VORTEX spinning machine" of Murata Machinery (MURATEC). The VORTEX spinning machine spins yarns in a unique way using an air vortex which was developed and is manufactured only by Murata Machinery.

Figure 6.14 A view of the Murata Vortex Spinning System

6.6 Principle of operation of MVS

The basic principle of operation is shown in Figure 6.15. The sliver is fed to 4-over-4 (or a four-pair) drafting unit. As the fibres come out of the front rollers, they are sucked into the spiral-shaped opening of the air jet nozzle. The nozzle provides a swirling air current which twists the fibres. A guide needle within the nozzle controls the movement of the fibres towards a hollow spindle. After the fibres have passed through the nozzle, they twine over the hollow spindle. The leading ends of the fibre bundle are drawn into the hollow spindle by the fibres of the preceding portion of the fibre bundle being twisted into a spun yarn. The finished yarn is then wound onto a package.

The drafting unit is also a very important element in Air-jet spinning. High drafts have to be performed, with good evenness of the fibre flow and excellent orientation of the fibres, at very high production speeds. Both suppliers have equipped the Air-jet machines with a 4- cylinder drafting system. As a comparison with the above-described wrapping mechanism with tandem jets, it is useful to consider the effectiveness of wrapping with only one jet. In this situation, the Z–twisting action of the jet is not nullified, and the core is therefore Z–twisted.

The edge fibres will, as described above, wrap the core with a Z–directional helix. Being that the core twisting is now in the same direction as the wrapping action, some edge fibre may become caught and twisted into the core, thereby reducing the number of them available to wrap the core.

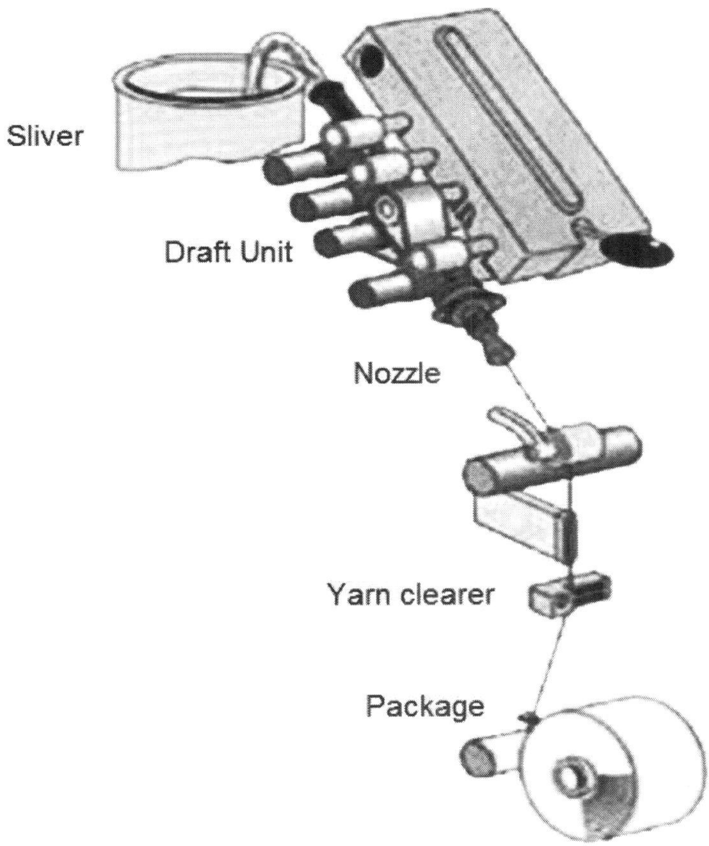

Figure 6.15 Principle of operation of MVS

One approach to increasing the number of edge fibres is to increase the length of the spinning triangle so that the twist insertion point is farther from the nip line of the front rollers. This results in edge fibres having to travel farther toward the twist insertion point at the apex of the spinning triangle. The effect of doing this can be demonstrated in conventional ring spinning by partially restricting the twist flow toward the nip line of the front drafting rollers. The result is an increase in edge fibres escaping twist insertion and becoming fly.

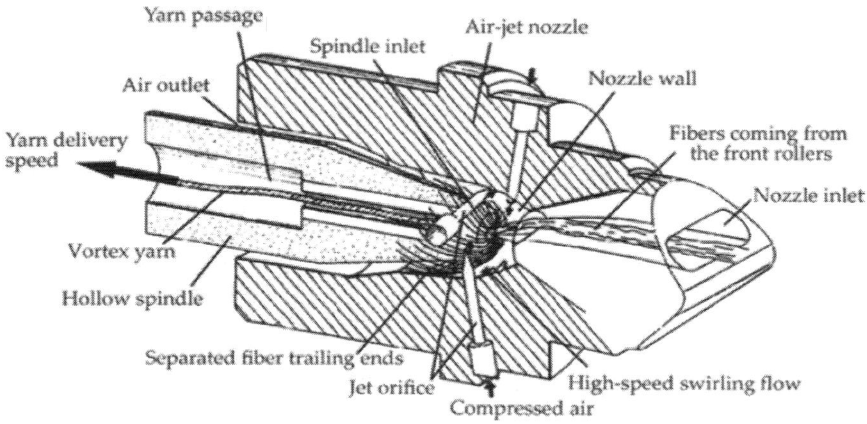

Figure 6.16 Schematic illustration of the nozzle block of vortex spinning machine

Figure 6.16 is a diagram of the single-jet design of the Murata Vortex system, which is a single air-jet spinning system. Compared with the tandem jet system, it incorporates a modified jet inlet. Little technical information is available on the working principle.

Much independent research has yet to be done to gain a detailed understanding of the wrapping mechanism of the Vortex system. However, it is possible that a partial blocking of the twist flow may occur above the jet nozzles to enable the formation of an extended spinning triangle and thereby increase the generation of edge fibres.

One problem with the vortex system is significant fibre loss during the yarn formation. This is related to the problem of variations in yarn quality which are not detectable by conventional evenness testers and sometimes only identified by weak points in the finished fabric. The path followed by the fibre in the currents created by the air jets play a crucial role in yarn quality.

Most structural defects are caused by the deflection of fibres in the air jet from their ideal path. Both delivery speed and yarn count are significant factors for yarn evenness and imperfections. An increase in delivery speed results in deterioration of yarn evenness. This is the result of decreasing efficiency of the air jet stream at higher delivery speeds because there is less time for the wrapper fibres to wrap around the parallel core properly. This particularly affects finer yarns and means that vortex spinning is best suited for coarser grades of yarn. Nozzle pressure also has a significant effect on yarn properties.

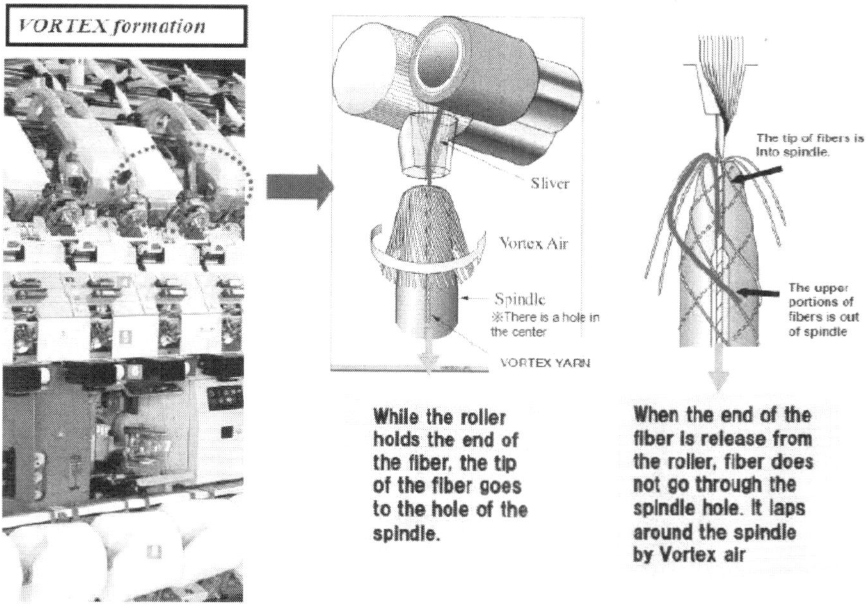

Figure 6.17 Schematic illustration of vortex formation

6.7 The structure of vortex yarn compared to other yarns

Vortex yarn has a two-part structure: a core surrounded by wrapper fibres. The number of wrapper fibres compared to the fibre core is higher compared to the air jet spinning. During yarn formation, the leading ends of the fibres are directed towards the yarn core and the trailing ends wrap around the core fibres. Such a structure provides the necessary fibre orientation and, at the same time, the required yarn strength.

This particular structure of the Air-vortex spun yarns influences the yarn properties, of course.

These properties are mainly determined by 2 parameters:

1. the percentage of wrapping fibres;
2. the twist level of the wrapping fibres.

The percentage of wrapping fibres can be influenced by the spinning draft and by the distance, and it is also a function of the yarn count.

Spinning nozzle

The spinning nozzle is basically the yarn formation element, i.e. the heart of the Air-jet spinning process. Compressed air at up to 0.6 Mpa enters the actual spinning chamber through 4 small bores, thus creating a very strong air vortex. At the outlets of the bores, this air vortex has a rotation speed of up to 1 000 000 rpm. The vortex performs 2 functions through this high speed:

1. generation of a vacuum and thereby an air flow through the fibre feed channel;
2. rotation of the free fibre ends around the spindle tip.

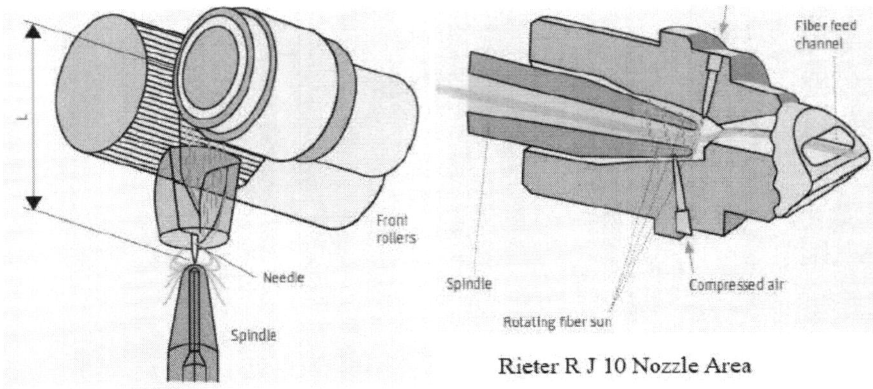

Rieter R J 10 Nozzle Area

Fibre Transport MVS

Figure 6.18 Comparison of MVS and RJ10 nozzles

A higher pressure can improve strength because wrapper fibres wrap more tightly around the core. However, it can also lead to more lost fibres. This creates potential weak points and increases unevenness in the yarn. A low pressure leads to improved evenness though strength is reduced. The distance between the front roller nip point and the tip of the spindle also affects yarn structure. The greater the distance, the higher the level of fibre wastage and yarn unevenness.

For generating free fibres ends, the correct choice of distance L (Fig.6.18) is very important. This distance should be slightly shorter than the average length of the fibres being processed. This enables the transport air in the fibres feed channel to separate fibre ends from the main fibre flow. It is evident that the longer the distance L, the more free fibre ends become available.

L is therefore an important process parameter. It is of course possible that during this process of fibre end separation, entire – mainly shorter – fibres are extracted from the main fibre flow.

These fibres have no chance of being integrated in the yarn. They bypass the spindle and are lost. In Air-jet spinning, the fibre loss (relatively short fibres) is therefore relatively high (5 to 10 %).

The higher the short fibre content in the sliver, the higher the ratio of fibre waste. By the action of the vortex, the fibre ends eventually whirl around the spindle tip and are thus twisted around the twistless yarn core and transformed into a twisted yarn surface or cover fibres. This occurs at the spindle tip. The twist of these surface fibres generates a certain torque in the yarn being formed. This torque has the tendency to twist the fibre bundle between drafting unit and spindle. Twist of this kind must be avoided in order not to interfere with the generation of the necessary free fibre ends. This can be solved by means of a twist stop. For this purpose Murata uses a needle (Fig.6.18), which detours the fibre bundle before entering the spindle, thereby acting as an efficient twist stop.

Once inside the spindle tip, the yarn formation process is finished, and the yarn can be taken off and wound onto a package. In Vortex yarns, the centre of the yarn is not twisted. Twisting occurs at the outer sides of the yarn, as shown in Figures 6.19 and 6.20. Fibres at the centre of the yarn remain loose while those at the outer side are fully twisted. In ring spun yarn, twist is given to the entire yarn from the centre to the surface of the yarn. The yarn thickness in vortex yarns is uneven. Twisting is concentrated at the thinner sections, while twisting is loose at the thicker section, leading to greater yarn hairiness. In the case of rotor yarn all the fibres are twisted from the centre to the outer side. Twisting is more uneven for fibres near the surface of the yarn. Table 6.1 provides a comparison between the three types of yarns.

As in two nozzle Air-jet spinning, the drafting unit is also a very important element in Air-jet spinning. High drafts have to be performed, with good evenness of the fibre flow and excellent orientation of the fibres, at very high production speeds. To achieve these goals, both suppliers have equipped the Air-jet machines with a 4-cylinder drafting system.

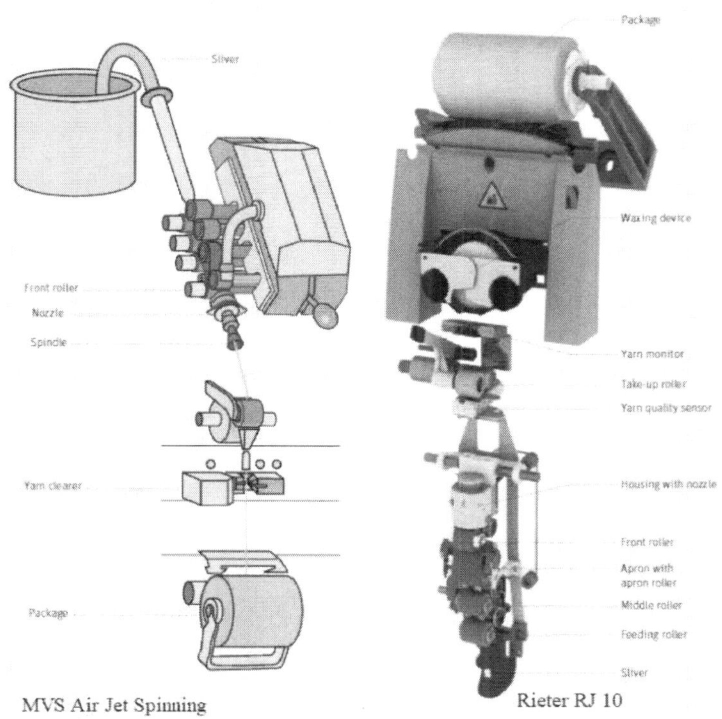

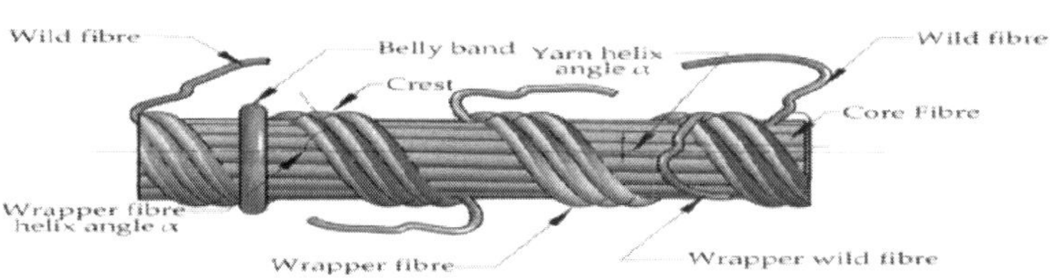

Figure 6.18 Schematic diagram of yarn fibre types in the fascinated yarn structure

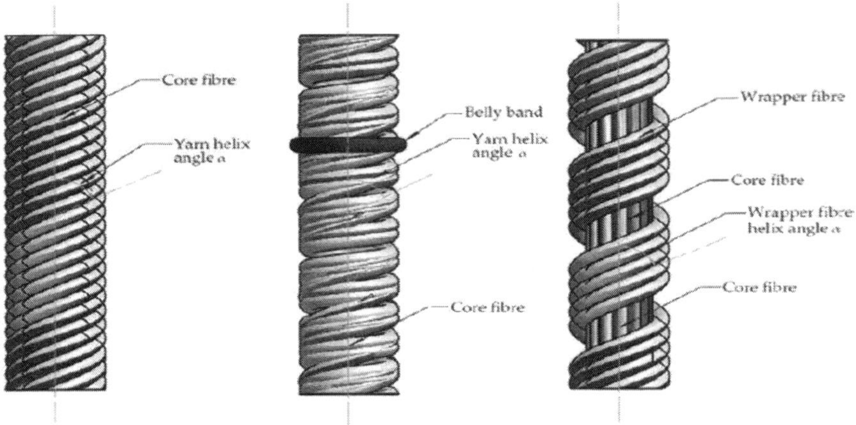

Figure 6.19 Respective ideal yarn structures of ring spun, Open-End and MVS yarn

Table 6.1: Comparison in yarn characteristics

Yarn Property	Ring Yarn	Rotor Yarn	Vortex Yarn
Parallelization and Fibre orientation	Max	Min	
Hairiness	Max		Min
Pilling	Max		Min
Strength	Max		Min
Unevenness	Max		Min
Surface smoothness	Max	Min	
Core fibres	Max		Min
Neps and Thick places	Min		Max
Bulkiness		Min	Max

The overall advantages of vortex yarn over the ring and rotor yarns are:

- Better resistance to pilling and abrasion: this gives longer-lasting fabric performance through a greater number of washing cycles.
- Less hairiness: this reduces potential problems in fabric production and gives a smooth appearance to the fabric.
- Less shrinkage: unlike ring spun yarn, the structure of vortex yarns means they are less prone to shrink
- Moisture absorption and drying: the looser structure of fibres at the Centre of vortex yarns means that they absorb moisture and dry quickly,

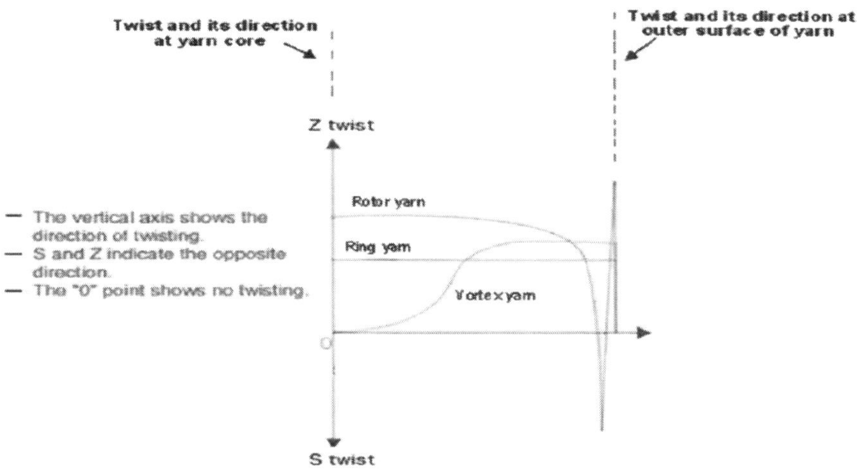

Figure 6.20: Difference in twist distribution in vortex, rotor and ring yarns

The figure below shows vertically the tightness of twisting, and Z-twist and S-twist indicate the directions of twisting centering on "0". The horizontal axis in the graph shows the distance from the center in a cross section of the yarn. The image chart of each type of yarn shows the yarn-twisting structure seen from the side.

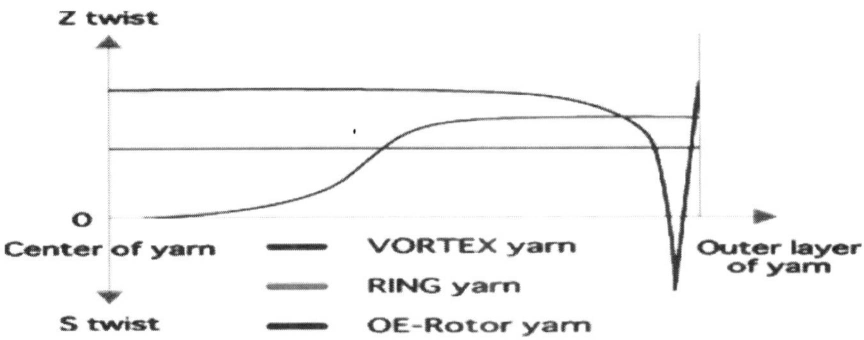

Figure 6.21: Difference in twist distribution in vortex, rotor and ring yarns

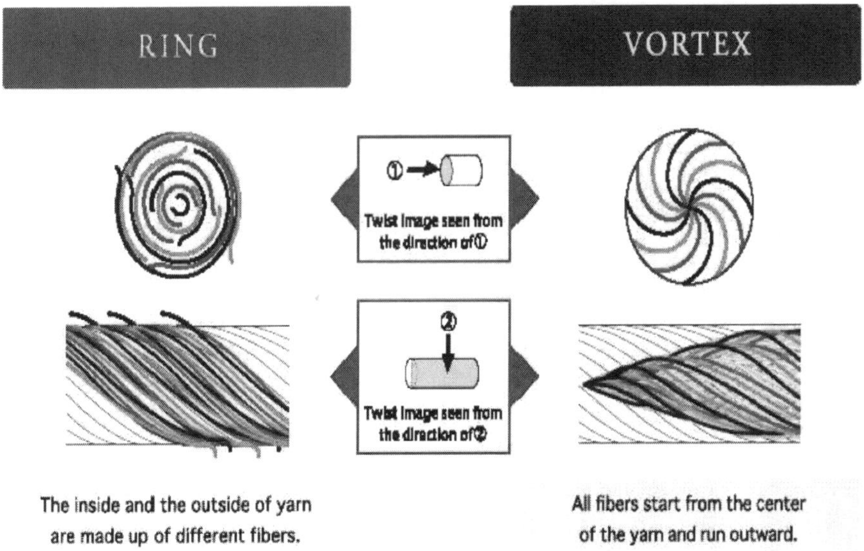

Figure 6.22 Comparison of ring and vortex yarns

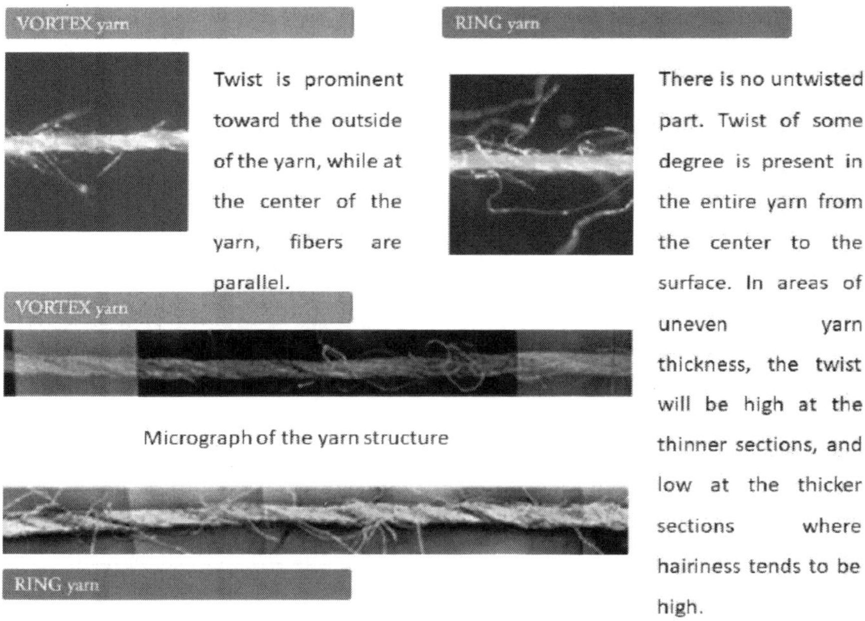

OE-Rotor yarn

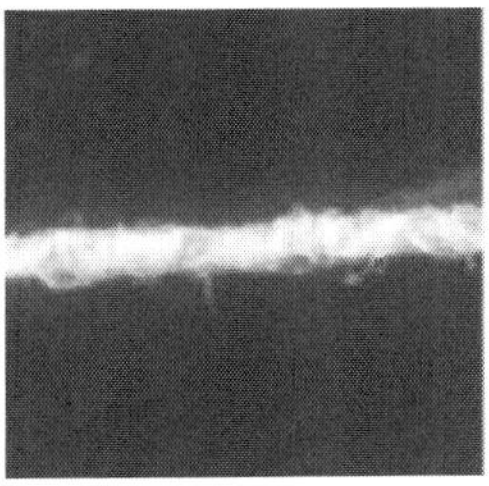

All fibers are twisted from the center to the outer side. For fibers near the surface of the yarn, twisting is uneven. Some fibers are twisted in the opposite direction of the main bundle.

Micrograph of the yarn structure

Figure 6.23 Comparison of ring, rotor and vortex yarns

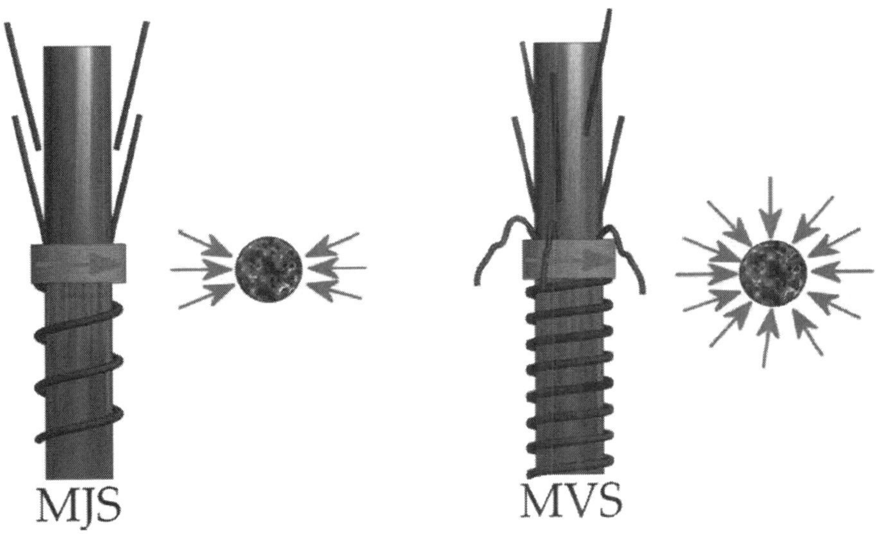

Figure 6.24 Comparison of MJS and MVS yarns

Chapter 7

Wool Spinning

7.1 Overview of early stage processing

There are two main systems used for the mechanical processing of wool from fibre into yarn: the worsted system and the woollen system (Figure 7.1). The semi-worsted system, which was developed as a high production spinning route for synthetic fibres, can also be used to process wool for carpet yarns. There are a number of processes common to both systems: blending of the wool types to be used; scouring of the wool to remove a range of impurities (e.g. dirt and grease); carding to disentangle the fibres and remove impurities like vegetable matter; spinning of fibres into a yarn. However, the technical aspects of these processes do differ between the systems due to differences in the 'raw materials' used by each. Also, there are additional processes specific to each system. The result is that worsted and woollen yarns differ in their structure and properties which ultimately influence the attributes of the fabrics produced from each system.

Processes such as shrink proofing, mothproofing, dyeing and finishing, for which there may be major differences between particular products, have not been shown for simplicity.

Approximately 80% of the wool clip enters the worsted system. It only uses virgin wools, generally finer than 27 micron, longer than 40 mm and with low levels of vegetable matter.

The worsted system utilises gilling and combing operations to give greater alignment of the fibres, resulting in a smooth and strong yarn. Both woven and knitted fabrics can be produced from these yarns, the fabrics being characterised by a smooth (or less hairy) surface. Men's suits and women's dress fabrics are examples of worsted garments.

7.2 Scouring

Scouring is an essential step in the conversion of wool into consumer products. The main objectives of scouring are to remove contamination and to minimise fibre damage in the process. Success in scouring is generally indicated by low levels of:

- contamination (i.e., residual grease, dirt, suint, vegetable matter and wool fragments)
- fibre breakage
- fibre entanglement and the achievement of good colour in the scoured product.

Traditionally the wool scouring process has been carried out in water, and today almost all wool is aqueous scoured rather than solvent scoured.

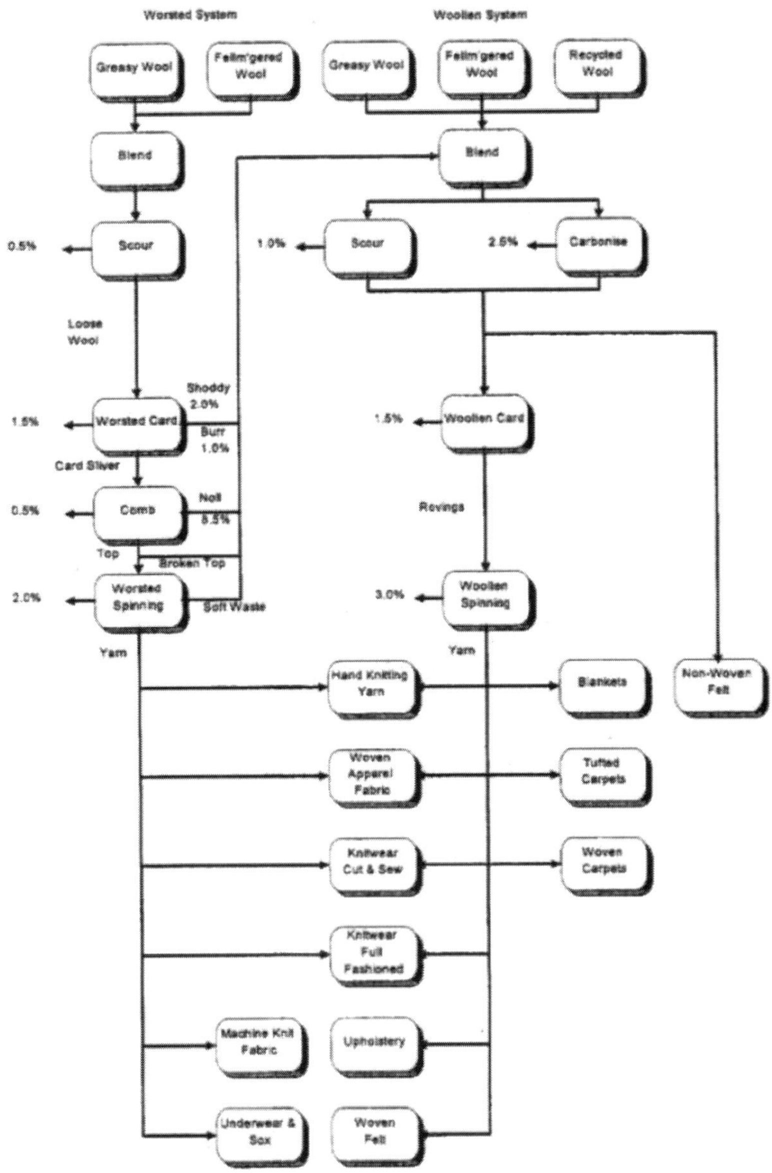

Figure 7.1 Basic processes in the worsted and woollen systems

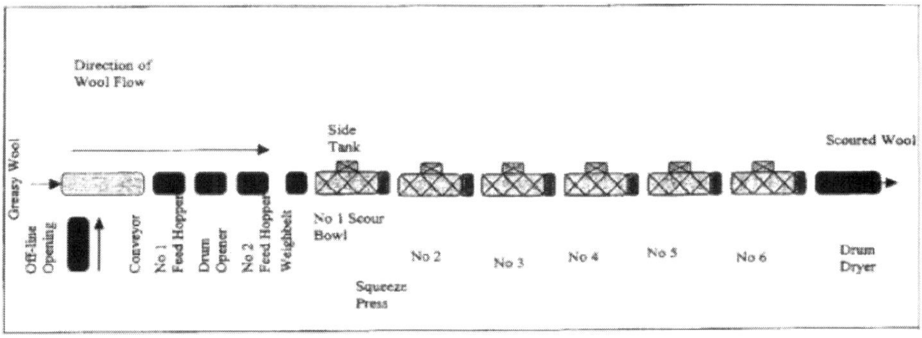

Figure 7.2 Typical worsted scour configuration

Figure 7.2 shows schematically the main components of a modern wool scouring plant designed to handle fine, apparel wools. The number of bowls and the number of hoppers in each bowl is influenced by the types of wool to be scoured. Care must also be taken not to over–scour very fine wools with treatments that are too long or too vigorous, otherwise excessive entanglement will ensure. Typically, the first two or three bowls will be scouring bowls, containing water and detergent at around 60-70oC. The final three bowls are rinse bowls, with water temperature generally around 40-50oC.

Associated with a wool scouring line are:

- ❖ Blending facilities, with machines for opening and dust removal;
- ❖ Heat recovery systems to transfer the heat from the effluent to the incoming water;
- ❖ Effluent treatment systems for separating wool grease from the effluent, and for ensuring that the discharged water has minimal contaminants;
- ❖ Hot air dryers to remove water and achieve the required level of regain;
- ❖ Baling machines to conveniently package the wool for shipment to a top making or a spinning plant.

7.3 Carbonising

Some fine wools contain high levels of vegetable matter (VM) of types which cannot be fully removed mechanically. Fragments of VM remaining in a yarn can cause serious problems for the apparel manufacturer. The only feasible way of ensuring that problematic VM types can be removed by mechanical action is to carbonise them first.

The carbonising process comprises a number of operations which must be matched to each other to provide a commercially acceptable product. Because of the chemical and physical processes involved (sulphuric acid, high temperatures and physical crushing), there must be a balance between:

- ❖ maximum removal of vegetable matter;
- ❖ minimal damage to the wool fibres (colour, strength and entanglement); and
- ❖ minimum fibre losses.

The following processing steps are involved in carbonising:

Acidising

The wool passes from the scour directly into one or more acid bowls where it is immersed in about 7% sulphuric acid.

Drying and baking

The wool is rapidly dried in a drum or conveyor dryer at temperatures below 70°C to prevent fibre damage. It then passes into the baking oven to dry and bake at around 120°C to 130°C. Here the vegetable matter becomes charred and brittle. A conveyor dryer is usually used for baking. Intermediate crushing and de-dusting may occur between the dryer and baking oven.

Crushing and de-dusting

The wool then passes through a number of sets of crusher rollers to break up the carbonized vegetable matter which should be brittle. These rollers operate under high force and care must be taken to limit fibre damage.

Neutralising

The dust–free wool then passes into the neutralising line which usually comprises about 4 or 5 scour bowls. Remaining sulphuric acid is neutralised with alkali such as soda ash and the wool is given a final wash and rinse before being dried as for normal scoured wool. The acidic nature of the carbonising process inevitably weakens the wool fibres and makes them unsuitable for worsted processing. However, carbonised wools may be used as part of blends for woolen processing.

7.4 Mechanical processing of wool

The mechanical processing stage in the wool pipeline commences with clean scoured wool and ends with the yarn ready for the fabric manufacturing stage. The main objectives of mechanical processing are to:

- disentangle and mix the fibres;
- remove vegetable matter;
- form a uniform strand of fibres (i.e., sliver or slubbing);
- attenuate the strand (i.e. reduce its thickness) and
- impart cohesion using twist to form a yarn of the required specifications and quality.

Other operations (which include some wet processing steps) include:

- lubrication of the fibres, to assist carding and spinning
- removal of short fibre and residual vegetable matter (combing in the worsted system only);
- dyeing (as loose stock, sliver, top or yarn);
- folding (or plying), winding, clearing,
- yarn scouring and twist setting (for carpets)
- shrink-proofing (loose stock or sliver).

All of the above processes must take place with maximum efficiency and quality, and with the minimum cost, fibre damage and breakage, and fibre loss.

The mechanical processing of wool can be divided into three main steps: sliver formation, sliver preparation and yarn formation.

Sliver or slubbing formation

This involves disentangling and mixing the fibres, removing vegetable matter and forming a continuous web, sliver or slubbing. This step is achieved by carding.

Preparing the carded sliver for spinning

This involves aligning the fibres (parallelisation), evening (doubling), drafting and the removal of short fibres, neps, vegetable matter and other contaminants. These are achieved by gilling, combing and drawing.

Yarn formation

This involves drafting the fibres into a thin strand and imparting cohesion to the strand (usually by inserting twist). This is the spinning step, which produces a singles yarn. Two or more singles yarns may be twisted together (plied) to form a folded yarn.

7.5 The three yarn manufacturing systems

The three alternative systems used for processing scoured wool into yarn are similar in many respects, such as the prior application of a special lubricant (as an emulsion with water) to assist processing, and the principles of the carding and spinning steps. The common features are shown in Figure 7.3.

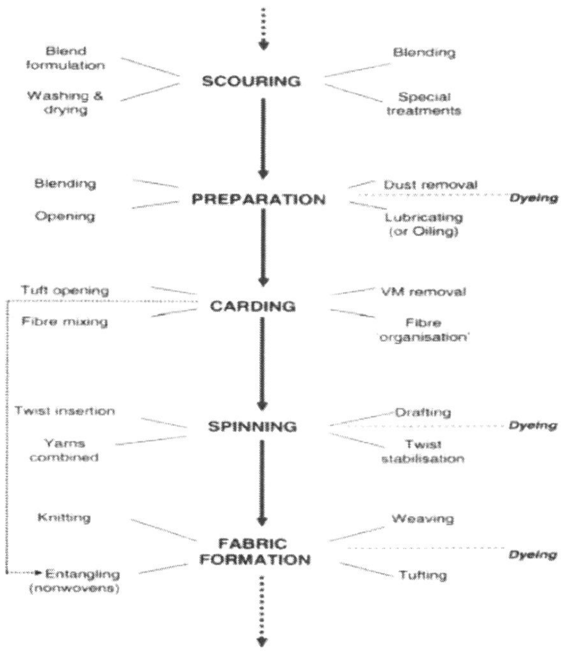

Figure 7.3 General steps in wool processing. Source: Wood, 2006

However, the three routes have significant differences, in terms of the number of steps required to produce a yarn, the type of machinery used at each step, the types of wools that can be economically processed, and the properties of the yarns produced (see Figure 7.4 and Table 7.1).

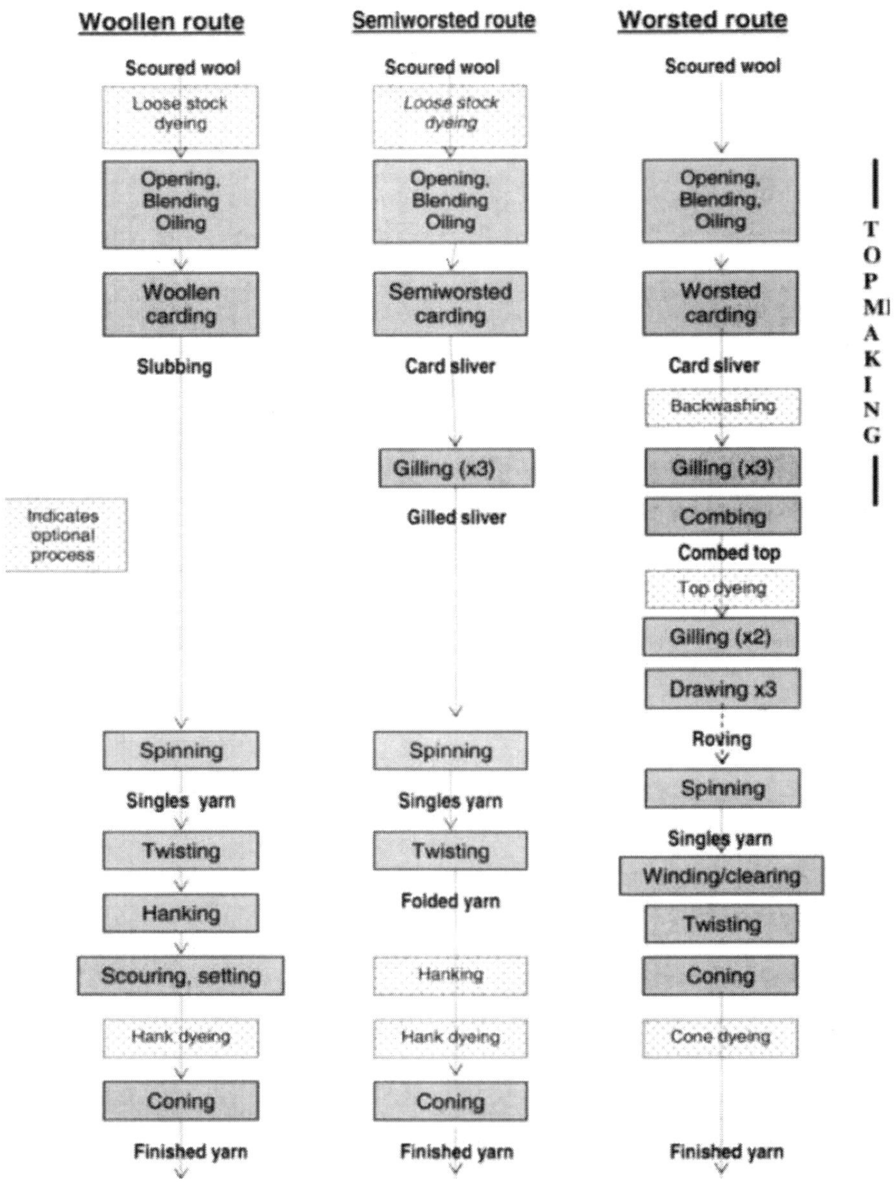

Figure 7.4 Steps in wool yarn manufacture. Source: Wood, 2006

Table 7.1 Comparison of the three processing routes. Source: Wood, 2006

	Woollen	Semi worsted	Worsted
Wool requirements:	Versatile; can handle all wool types, but more suitable for shorter wools rather than very long	Wools should be sound, staple length 75-125 mm, and low vegetable matter content	Requires wools which for their diameter are longer, better style and sounder
	Usually a wide range of blend components	Usually a limited range of components in blend	Similar wools used rather than a mixture of types
	All fibre diameters used, from very fine to very coarse wools	Mainly medium fineness wools; 27 - 35 µm	Mainly fine wools; less than 30 µm, usually less than 24 µm
	Can use reprocessed wools of all types	Not suitable for short reprocessed wools	Waste wools never used
	Blend cost generally lowest	Blend cost higher than woollen	Blend cost highest
Complexity of the processing system	Shortest route with fewest steps, large woollen card has low production rate	Compact, high production system, cheaper to operate than worsted system	Most complex route; largest number of steps; card similar to semi-worsted card
	Card removes some VM but cannot tolerate high levels of vegetable matter	Limited ability to remove vegetable matter and short fibres	Vegetable matter and short fibres removed by combing
	Carding is very critical because it sets the yarn count, and is the final opportunity for blending	Carding is less critical because of substantial blending and drafting in subsequent steps	Carding is less critical because of substantial blending and drafting in subsequent steps
Properties of yarn:	Minimal alignment of fibres, many may be hooked	Reasonable degree of fibre alignment	Fibre alignment very high, giving most even yarn
	Hairy yarn - many fibres protrude from surface	Fewer fibre ends and loops protrude from surface	Few fibre ends and loops protrude, so least hairy
	Yarn is bulky, soft and resilient	Medium bulk and resilience	Low bulk (lean, smooth), and firm handle
	Tends to be weakest, with breaking strength of 3-5 g/tex	Stronger, with typical breaking strength 5-7 g/tex	Tends to be strongest yarn; breaking strength 7-9 g/tex
	Requires at least around 110 fibres in yarn cross-section	Requires at least around 90 fibres in yarn cross-section	Requires at least around 40 fibres in yarn cross-section
End product uses:	Suitable for all purposes: apparel, carpets, furnishings.	Carpets and knitwear mainly	Fine weaving and knitting yarns for quality apparel
	Yarn structure often not apparent in finished product	Yarn structure may or may not be clearly visible	Yarn structure is usually well-defined in product

Fibre preparation for spinning

Before yarn manufacture can commence, some preparation is required in the spinning or topmaking plant. Preparation involves opening, blending and lubricating, processes which prepare the wool for spinning by:

- separating the scoured wool into smaller clumps;
- removing as much dust and short fibre as possible by vigorous agitation;
- combining different types of wool and other fibres into the processing blend and thoroughly mixing them;
- spraying a light film of lubricant (oil and water) on the wool, to assist processing.

A common opening machine is a Fearnought, shown in Figure 7.5. It operates rather like a card, with rotating toothed rollers that interact to reduce the size of the clumps of scoured wool.

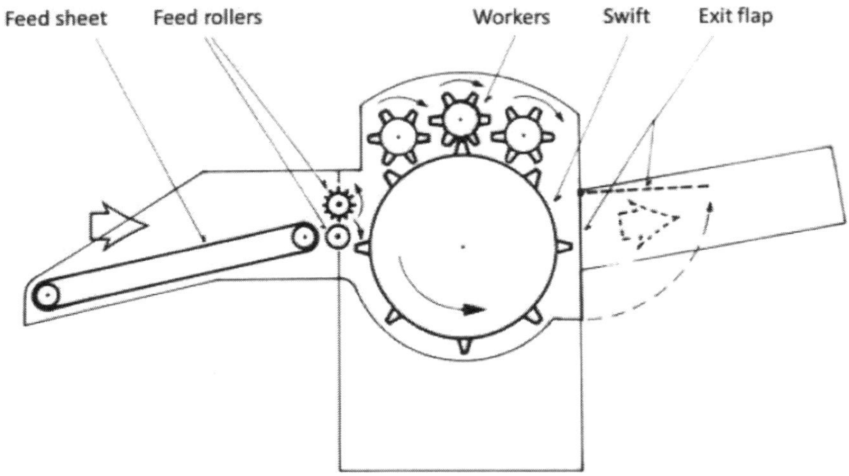

Figure 7.5 Fearnought wool opener. Source: Wood. 2006

While all three routes require these processes to be carried out, a higher level of lubricant (~3%) is usually applied for woollen spun yarns than the other routes to promote sufficient fibre cohesion. Furthermore, woollen blends tend to require more thorough blending because:

1. A wider variety of wool types is generally used, and
2. The woollen route provides no significant opportunities for blending after the carding step.

Figure 7.6 shows a typical blending system for woollen yarn manufacture.

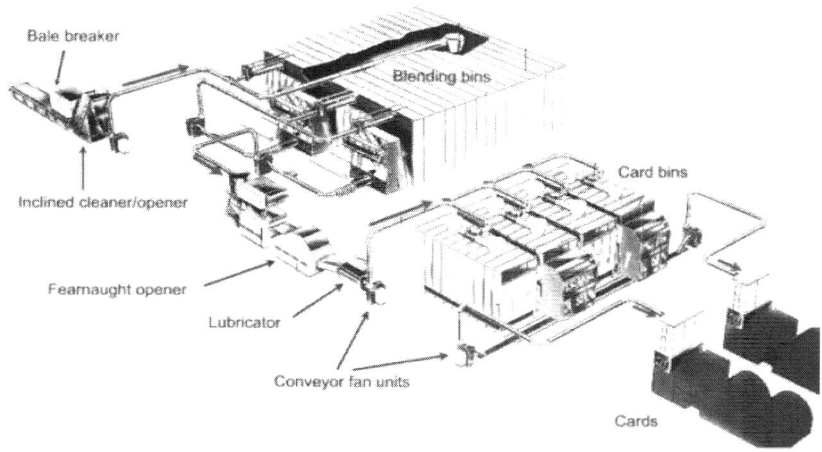

Figure 7.6 Typical blending system for woollen yarns. Source: Wood, 2010

Worsted system

The worsted system has the most steps of the three routes. Better-style, sound wools are required to ensure efficient processing and acceptable yarn quality. There should not be any defects in the wool, such as tenderness or cotts, or too much vegetable matter. Fine and medium wools are preferred to produce a fine, lean, flexible yarn, and ultimately a light, soft-handling fabric.

Carding

After the scoured wool is opened, blended and a lubricant applied, carding is carried out. This is the process where metal teeth or pins (on rollers rotating at different speeds) tease the tufts of fibres apart and lay the fibres roughly parallel into a card sliver. Carding also removes some of the vegetable matter present in the wool. . The wool is removed from a worsted card and coiled into a can as a thick ribbon of fibre, called a sliver.

Figure 7.7 shows a typical arrangement of the rollers in a worsted card. A semi-worsted card is very similar to a worsted card in construction, but generally runs at a higher production rate.

Gilling

More extensive fibre organisation steps are required in the worsted system, so the card sliver is passed through a series of gilling steps to straighten and align fibres in a neat, parallel arrangement. The action of a gillbox (Figure 1.8) closely resembles the combing of hair. Gilling involves lines of metal teeth, mounted on a series of steel bars (fallers), being drawn through the sliver as it moves through the machine.

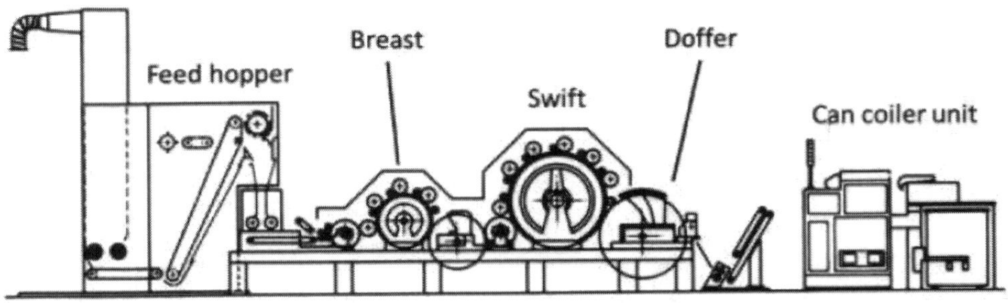

Figure 7.7 Worsted card. Source: Wood, 2010

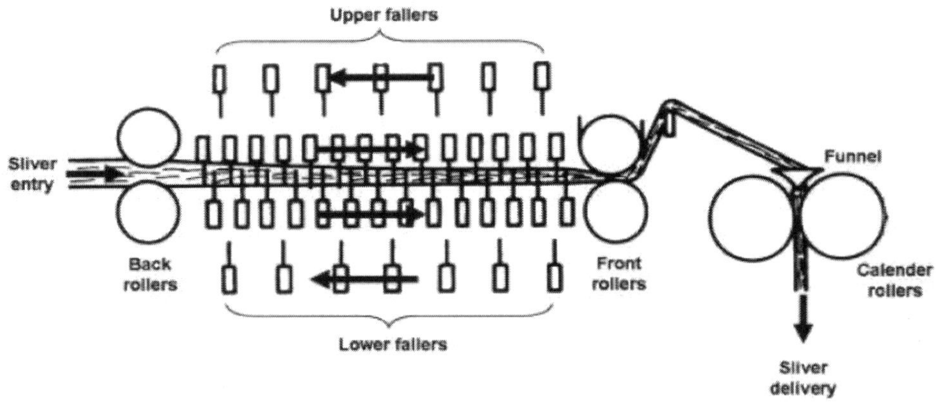

Figure 7.8 Principles of gillbox operation. Source: Wood, 2007.

Combing

The gilled sliver is combed to remove short fibres (noils), neps (tiny clumps of fibre) and vegetable matter (Figure 7.9), then gilled again to restore the parallel alignment to form a top. An optional backwashing treatment may be included at this stage, where the top is scoured to provide a whiter, cleaner product.

The top is drafted (i.e. drawn out to reduce its thickness) to form a thin, uniform ribbon o fibres, called a roving (Figure 7.10), before being further drafted and twisted in spinning to form a singles yarn (Figure 7.11). In the worsted system the first group of processes, i.e. blending, carding, gilling and combing is called topmaking.

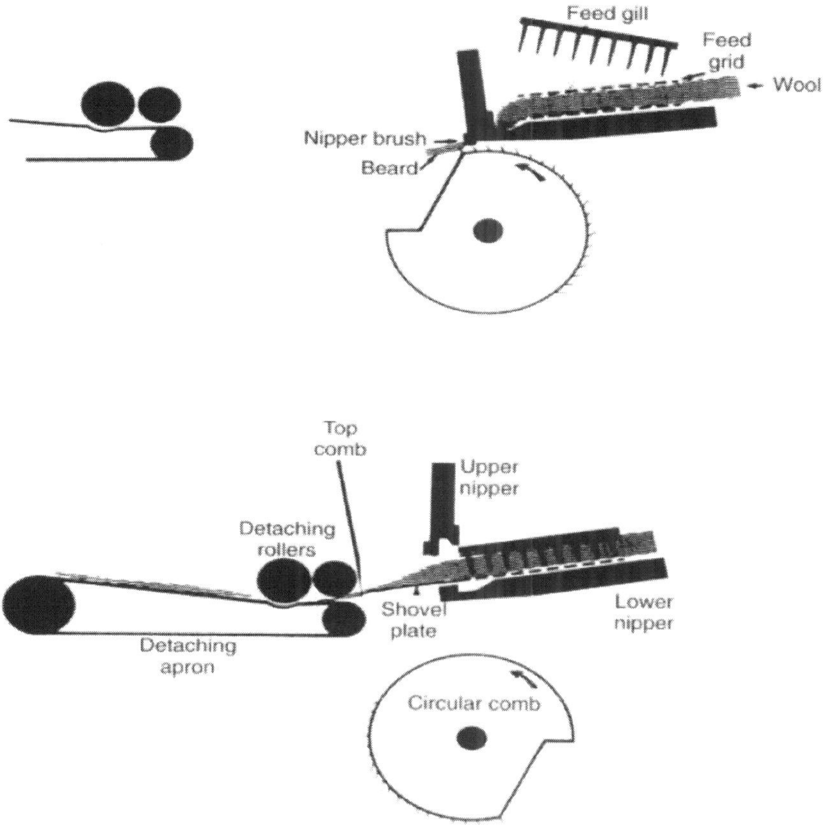

Figure 7.9 Combing action. Source: Wood, 2007.

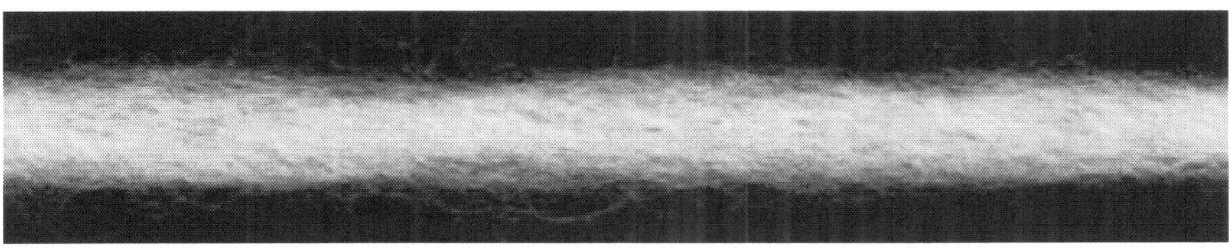

Figure 7.10 Roving. Source: Wood, 2006.

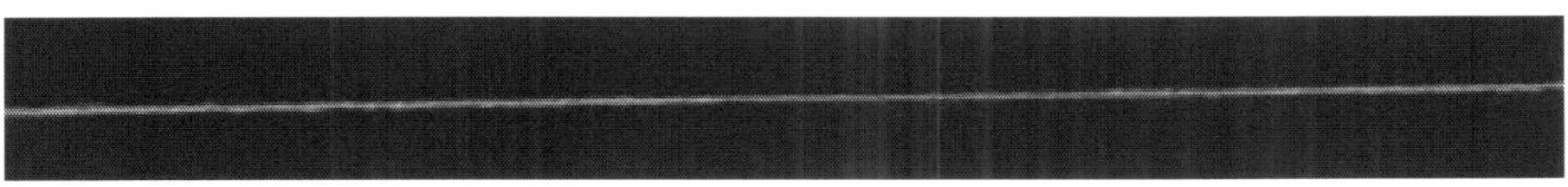

Figure 7.11 Worsted yarn. Source: Wood, 2006.

Drafting

In the production of a worsted yarn, the wool goes through a number of intermediate stages to gradually reduce its linear density (or thickness). A top may have 25,000 fibres in its crosssection while a worsted yarn may have only 50 fibres or less. To achieve the required yarn fineness, a considerable reduction in the thickness of the structures must occur, through successive drafting at each stage. The sequence of fibre structures in the worsted system is:

scoured wool – card sliver – gilled sliver – top – roving – singles yarn – folded yarn Roller drafting (or drawing) is carried by passing a sliver or roving between two pairs of driven rollers.

The delivery (or front) rollers have a higher surface speed than the feed (or back) rollers. For example, if the surface speed of the front rollers is ten times the surface speed of the back roller (i.e. draft ratio = 10), there will be a ten-fold reduction in the thickness of the sliver. Thus a thinner strand will emerge from the drafting zone.

Spinning

Figure 7.12 shows the spindles on a worsted ring spinning frame, which is the most commonly used machine for spinning worsted yarns. The rovings (top of picture) pass through the drafting zone (middle) and the yarn is wound onto the bobbins (bottom), which rotate at high speed on spindles.

Figure 7.12 Worsted spinning frame. Source: Wood, 2006.

As a result of (1) the removal of short fibres in combing, (2) the parallel arrangement of the fibres presented for spinning, and (3) the high degree of twist imposed, a worsted yarn is sufficiently strong to permit a minimum of around 40 fibres in the cross-section. Hence fine, even, firm yarns with sufficient strength for weaving or knitting can be spun by the worsted process. Figure 7.13 shows the regular helical path of a typical fibre, imposed by the twist in a worsted yarn.

Figure 7.13 Helical path of a fibre in an ideal worsted yarn. Source: Wood, 2006.

After spinning, two (or more) singles yarns may be twisted together to form a plied yarn. At the same time the yarn may be cleared, i.e., have faults such as thick or thin places removed and the yarn re-joined automatically.

Characteristics of worsted yarns

Worsted-spun yarns are mostly used in high-quality woven suiting fabrics, and hand and machine knitting. In these products it is important that the yarn be free from faults such as neps, slubs (lumps), vegetable matter and short, protruding fibres. Short fibres can cause discomfort in wear due to the prickle sensation. Long, protruding fibres have a propensity to entangle and form pills (small clusters of fibres on a fabric surface), which are visually unattractive.

Woollen system

The woollen system is the simplest of the three yarn manufacturing routes, and because it is the most versatile, it tends to be used to process blends of wools which vary in length and fibre diameter. The system is capable of handling the poorer-style wools and the short and tender wools. Noils (combing waste) and recycled fibres (e.g. combing noils and from rags and yarns) may also be included in woollen blends. Crossbred wool blends destined for carpets are mostly processed on the woollen system.

Woollen carding

A woollen card is, by necessity, a more complicated machine than a worsted card because it provides the final opportunity for fibre mixing before spinning, and it also determines the count (or thickness) of the yarn. Instead of producing a sliver (as in the worsted and semi worsted cards) it forms slubbings using a device called a condenser. A slubbing is similar to a worsted roving, but it is less uniform and the fibres in the strand are much less well-aligned and straight.

Any variations or irregularities in the slubbings persist through spinning and hence may affect the quality (i.e. evenness) of the finished yarn. A woollen card may be up to 3.5 metres wide and 15 metres or more in length.

From the hopper feed the wool generally passes through two carding stages (termed the scribbler and carder parts respectively) to ensure a thorough opening of the wool tufts and the mixing of fibres, and the removal of contaminants (Figure 1.14). A web-purifier (or Peralta) is a pair of smooth steel rollers that crushes the vegetable matter into fragments, hence making them more easily removed from the web. The intermediate feed conveys the card web from the scribbler section to the carder section.

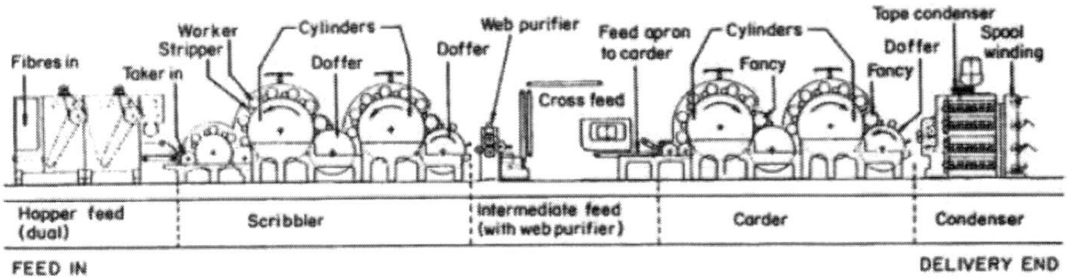

Figure 7.14 Layout of a woollen card. Source: Wood, 2006

At the tape condenser (Figure 7.15), the web (i.e., a thin fibrous film) of carded fibres is separated into narrow strips which are consolidated by a rubbing action into slubbings. The condenser tapes convey the strips of fibres into the rubbing aprons and the slubbings formed there are wound onto spools (or cheeses) ready for spinning.

Woollen spinning

The slubbing produced by the card is spun into yarn by applying twist to it. This step is most commonly carried out by a machine called a ring spinning frame (Figure 7.16). A ring frame consists of a series of spindles which rotate at high speed (around 6000 rpm or higher) onto which the yarns are wound. The yarn is guided onto the tube by a small clip called a traveller, which is drawn at high speed around a metal ring by the rotation of the spindle. At the same time, the ring rail oscillates slowly in the vertical direction to enable a tidy, compact package of yarn to be formed on each tube.

Unlike worsted and semi-worsted spinning, the shorter blends mostly used in the woolen system are not suited to roller drafting. Strands of short, poorly-aligned fibres do not draft as well as longer, straighter fibres. Hence, the draft used on a woollen frame is minimal (i.e. a draft ratio ~ 1.3 compared with 20-30 in worsted spinning and up to 50 in semi-worsted spinning).

A false twist device between the front and back rollers inserts a temporary twist in the slubbing. This provides strength for drafting as well as assisting in producing a more even yarn. The thicker and less twisted sections draft more than thinner, more tightly twisted sections. The sequence of fibre structures formed in the woollen system is: scoured wool – slubbing – singles yarn – folded yarn

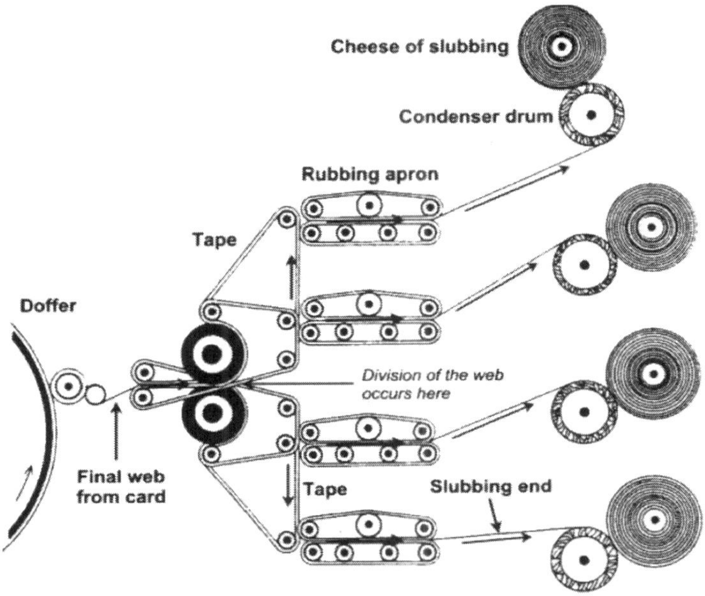

Figure 7.15 – Parts of the condenser on a woollen card. Source Wood, 2006.

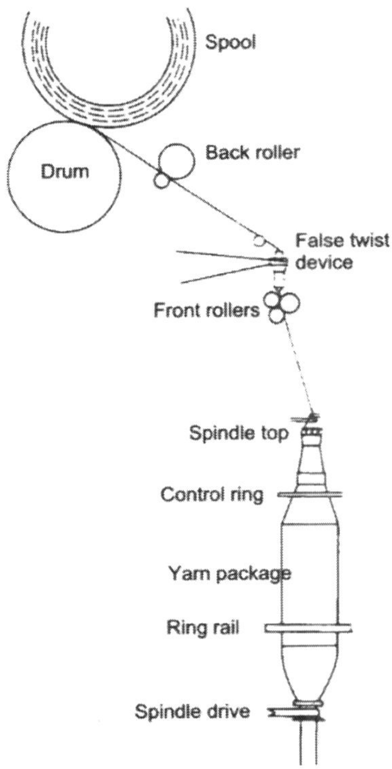

Figure 7.16 Key parts of a woollen spinning frame. Source: Wood, 2010.

Mule spinning

Ring spinning, which is the most common method of producing a woollen spun yarn, is a continuous process, i.e., the steps of drafting, twist insertion and winding onto a package operate simultaneously without interruption. The alternative method of producing a woolen yarn, mule spinning (Figure 7.17), is an intermittent process. Here, the drafting and twist insertion steps alternate with the winding-on step.

While mule spinning frames have slower production rates than ring frames, use smaller packages and occupy more floor space this method of spinning is undergoing a small resurgence. This is because, in comparison with ring spinning, mule spinning is said to give a more even, finer yarn, and better quality yarns from shorter blends, such as those containing lamb's wool, possum and cashmere fibres.

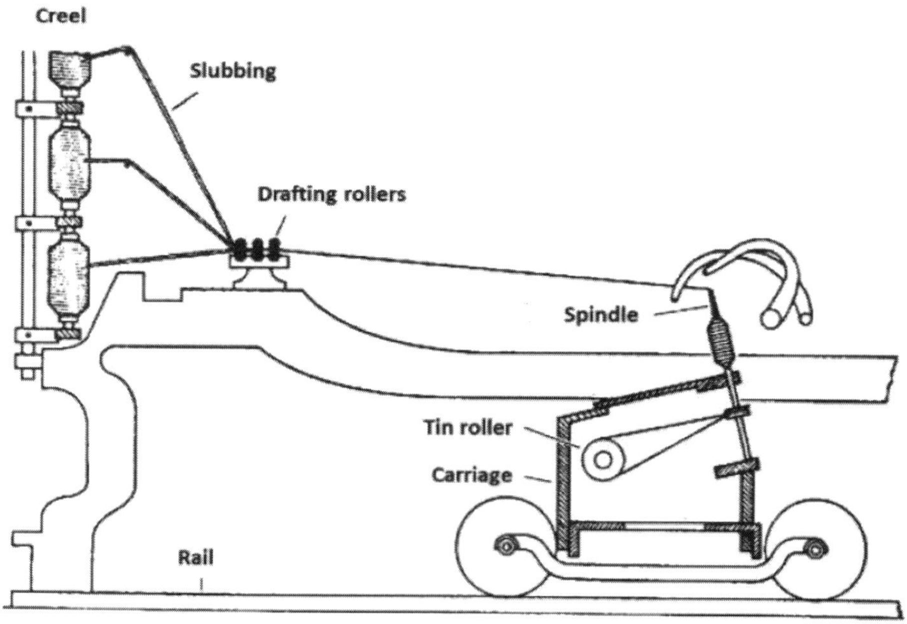

Figure 7.17 Mule spinning frame. Source E J Wood, 2010.

Other steps in the woollen system

Lubrication

In conventional woollen spinning, a processing lubricant, an emulsion of oil with water, is sprayed as a mist onto the wool before carding. (The usual rate of application of about 3% on weight of fibre is significantly higher than for worsted and semiworsted spinning where levels around 0.3% are common.) The lubrication treatment has three purposes:

- Increasing the moisture content of the wool fibres, making them more extensible, hence more resistant to the stresses of carding, and also reduces electrostatic effects
- Reducing the friction between fibres, and between fibre and the card teeth, thereby

reducing fibre breakage in carding,

- ❖ Improving web cohesion, so fewer fibres are lost as droppings or fly.

Twisting (or plying)

To provide sufficient strength and improve the uniformity of woollen yarns, and reduce the tendency to untwist, two or more singles yarns produced are usually twisted together to form a folded or plied yarn. It is usual to ply worsted and semiworsted yarns too. The methods used for wool yarns are ring twisting and two-for-one twisting.

Yarn scouring and twist setting

To avoid downstream problems such as excessive soiling of the yarn, the processing lubricant must be removed from the yarn in a process that is similar to wool scouring. At the same time, for yarns destined for cut-pile carpets, one bowl will contain the chemicals required for a setting treatment that enables the yarn to resist untwisting when cut. Insect resist agents can also be applied to the wool during this process. If scouring is not required a wool yarn can be set by steam in a continuous autoclave such as the Superba system.

Characteristics of woollen spun yarn

A woollen-spun yarn is characterised by a high proportion of short fibres distributed very much at random throughout the yarn. Longer fibres often undergo reversals in direction and may protrude from the yarn, to contribute to a hairy and quite irregular appearance, as shown in Figures 7.18 and 7.19.

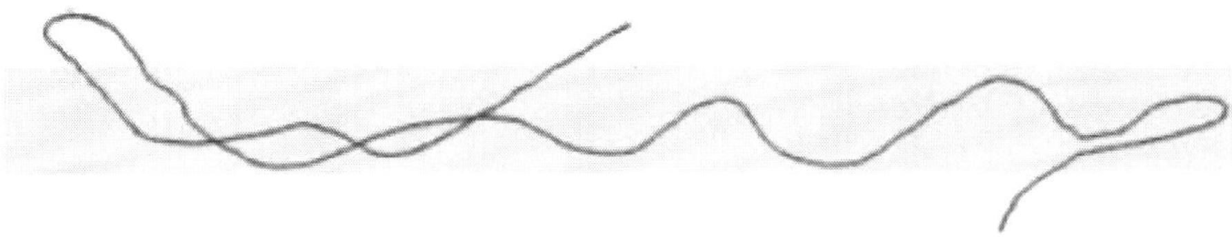

Figure 7.18 Typical fibre path in woollen yarn. Source: Wood, 2006

Figure 7.19 Two-fold woollen spun carpet yarn.

The loops and ends of fibres protruding from the surface have an important influence on the tactile and visual properties of a woollen-spun yarn. A fibre with reversals contributes less to the strength of a yarn than if it was fully extended without reversals, as it would be in a worsted yarn. Therefore, woollen yarns tend to have less strength than worsted yarns, and to obtain sufficient strength a minimum of around 120 fibres in the cross-section is required. The yarns produced by this route tend to be of a coarse count (i.e. thicker), spun to a low twist, and hence are bulky, soft handling, hairy, and less regular than worsted yarns. The reversals and other irregularities in the paths of the fibres within the yarn create air spaces, which also contribute to the soft, bulky handle of woollen-spun yarns (provided the twist level is not high). In the main, woollen spun yarns are used in woven, knitted and tufted products such as carpets, blankets, tweeds and heavier woven and knitted apparel.

The semi-worsted system

This system was originally developed to produce a yarn with greater strength than in the woollen system, but without the expense of removing short fibres in a combing step. Wool is passed through a high-production card, gilled several times to straighten the fibres, and spun directly from a thin sliver. In the semi worsted processing route the sequence of structures is:

scoured wool – card sliver – gilled sliver – singles yarn – folded yarn

,Because of the need for superior fibre length for the high drafts often employed in spinning, the semi-worsted system requires sound wools of 100-120 mm staple length (with a minimum length of at least 70 mm). There is less opportunity for the removal of contaminants, so wools should be free of vegetable matter.

Semi-worsted yarns are intermediate in properties between those produced on the worsted and woollen systems. In New Zealand they are used mainly for manufacturing carpets and knitwear, but in other countries the semi worsted route is a high production yarn-making process, used mainly for synthetic staple fibres. Because of the greater fibre length and the degree of straightening introduced by the gilling step, semi-worsted yarns tend to be less bulky than woollen yarns. Hence bulkier wools are sought as a major component for a semi worsted blend if good fabric cover or a softer handle is required in products. However, the quite lean, lustrous appearance of semi-worsted carpet wool yarns means that their use is restricted to relatively dense, loop pile constructions, and they are rarely used in cut-pile products.

Summary

To make almost all wool products the wool must first be scoured and impurities such as vegetable matter removed. The fibres are disentangled from the clumps of wool by carding, made parallel and then spun into yarn (unless intended for nonwovens). Finally the yarns are interlaced by weaving or knitting to form a fabric. The type of processing route, and the range of products that can be made, depends largely on the properties and quality of the wool. If the wool is heavily contaminated with vegetable matter, carbonising may be required to assist in its removal.

Three alternative yarn manufacturing sequences are used for wool; i.e., the woollen, worsted and semi-worsted routes. This topic briefly describes and compares the three routes, the steps involved and the types of yarn generally produced by each route. The specific processing steps are examined in more detail in subsequent topics.

Chapter 8

Filament Yarn Spinning

8.1 Fibre-Extrusion Spinning

There are various fibre-extrusion spinning methods for the manufacture of synthetic fibres, but the most commercially popular is melt-spinning for thermoplastic polymers. It is followed by wet spinning (solution wet spinning) and dry spinning (solution dry spinning), which are used mainly for non-thermoplastic polymers.

8.1.1 Melt-Spinning

Melt spinning can be defined as a continuous process in which a thermoplastic polymer is heated to form a molten solution that is extruded by pumping it through an array of very small orifices to form molten filaments. These are subsequently cooled to their solid state while being simultaneously attenuated by stretching, thereby producing what is termed as 'as-spun' or 'partially drawn' CF yarn. With the exception of PP, all of the thermoplastic polymers have some degree of moisture absorbency and therefore the polymer chips have to be dried prior to melt spinning. Figure 8.1 shows the basic vertical arrangement of the melt-spinning process, called the spinning line.

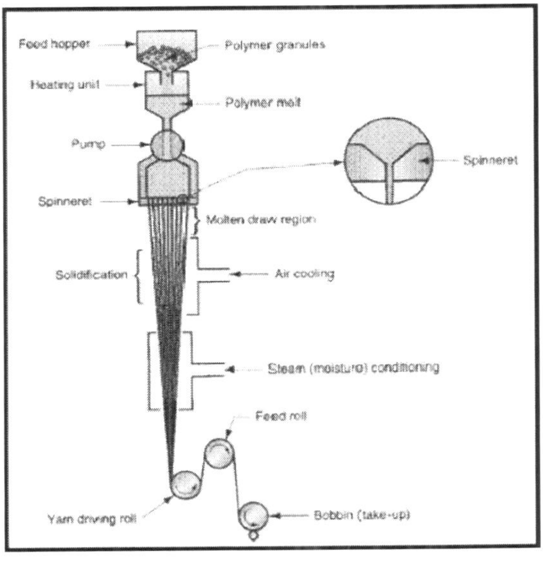

Figure 8.1 Schematic illustration of melt spinning line (Source: Engg. Encyclopedia)

The dried polymer chips or pellets are supplied to a screw extruder via a hopper. The screw extruder has several hot zones to progressively heat the polymer chips to a few degrees above their melting point, Tm (melt temperature). The molten solution is fed to the metring pump and it also kept at a temperature above the Tm. The pump is connected to a spin pack (Figure 8.2).

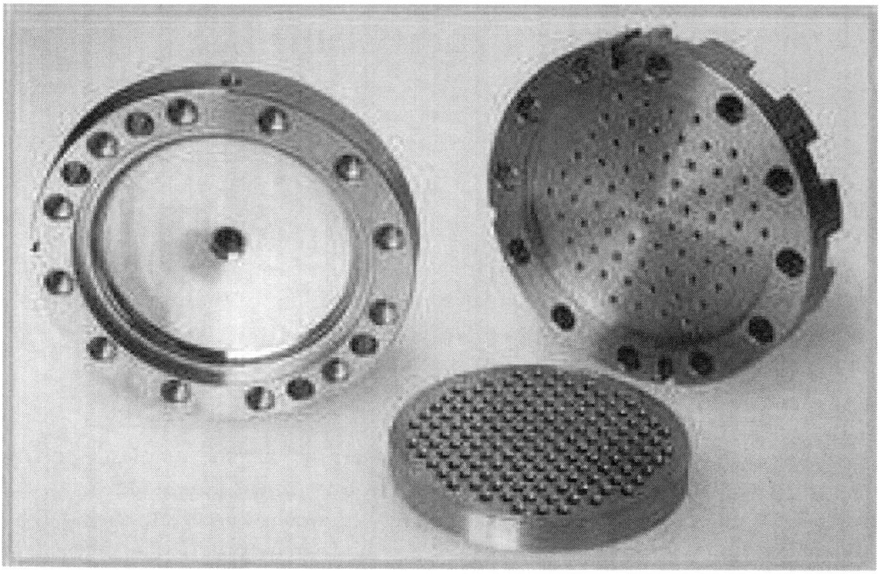

Figure 8.2 Spin pack and spinneret

This comprises a stack of circular plates with metal filters sandwiched between them. Each plate, subsequent to the inlet plate, has a series of holes through which the polymer flows until it reaches the last plate, the spinneret. The spinnerets may have between 500 and 4000 holes per disk. The orifices in the spinneret largely govern the diameter of the molten filament streams; these are stretched while being cooled to their solid form. Stretching occurs as the partially solidified, semi-molten filaments are pulled down from the spinneret holes at a faster speed than the molten flow through the orifices. The ratio of the former to the latter speed is called the draw-down ratio. The pump is designed to maintain a constant flow through the holes of the spinneret, hence the term 'metring pump'.

This enables a relatively consistent molten filament diameter. The resulting filament diameter is therefore largely dependent on the stretching by the godet rollers, i.e. the draw-down ratio. The Tm of the polymer and the flow of the molten polymer are important factors in the process.

The ease of flow is directly related to the viscosity of the polymer, which in turn is a function of its molecular weight. Polymers suitable for textile fibre production have high molecular weights, and therefore these molten polymer solutions have high viscosities, which means they do not easily flow.

However, designed to maintain a constant flow through the holes of the spinneret, hence the term 'metring pump'. This enables a relatively consistent molten filament diameter. The resulting filament diameter is therefore largely dependent on the stretching by the godet rollers, i.e. the draw-down ratio.

The Tm of the polymer and the flow of the molten polymer are important factors in the process. The ease of flow is directly related to the viscosity of the polymer, which in turn is a function of its molecular weight. Polymers suitable for textile fibre production have high molecular weights, and therefore these molten polymer solutions have high viscosities, which means they do not easily flow. However, viscosity decreases with increasing temperature and consequently extrusion commonly occurs at temperatures higher than the Tm. For example, with PET, extrusion temperatures (spinning temps, Ts) may be within 280–290 °C for molecular weights within 15,000–20,000. Table 8.1 summarises typical spinning temperatures for the major synthetic polymers along with their melting points.

Table 8.1 P roces s temperatures for synthetic polymers

Temperatures (°C)	PET	PBT	PTT	PLA	PA 6	PA 6.6	PP	PU
Melting point (*Tm*)	265	225	228	150–160	220	265	168	280
Glass transition point (*Tg*)	80	66	45–65	55–60	40–87	50–90	−17/−4	−38/−35
Spinning (*Ts*)	280–300	250–280	240–270	220–240	240	280	220	300

The study of the melt-flow behaviour is referred to as the rheology of the molten liquid, and enables determination of the pump pressure required for a steady flow of material through the spinneret. Pump pressures can be up to 70 MPa2.

The ratio of length to diameter of the spinneret holes is also important to achieve a steady flow. This can range from 2 to 5, depending on the polymer. The structure and properties of the spun yarn are strongly influenced by the cooling and attenuation of the extruded filaments in the spinning line. The action of attenuation by stretching is called drawing. Drawing tends to straighten and align the constituent polymer chains with the fibre axis; this improves the tensile properties of the filaments. Increasing the speed of draw-down godet rollers therefore increases the polymer orientation. Depending upon the way in which the melt-spinning process is implemented, the filaments may be subjected to low, moderate or high draw-down ratios. When low and moderate ratios are used, a second stage of drawing is applied by reheating the filaments whilst stretching them. The filaments are reheated to just above the temperature at which they are still in a solid state but very easily deformable (plastic state). This is called the glass transition temperature (Tg) (see Table 8.1). The filaments that result from this stage are referred to as fully drawn. Although drawing directly induces orientation of the polymer chains of the filaments, because of chain entanglement, not all the lengths of the polymer chains will become aligned and parallel with the filament axis. A fully drawn filament therefore, will be orientated to the limit that may be obtained with the particular polymer. Fully dawn filaments can be achieved in one of two ways. Heated godets may be fitted to the melts pinning machine and positioned after the pull-down godet. Each godet is set to rotate at a faster surface speed than the previous one. While the filaments are being heated, the speed differences impart the stretching or drawing and the ratio of the faster to the slower speed gives the draw-ratio. It is usual for hot drawing to be carried out progressively so a sequence of godets is employed, the number depending on the total draw ratio required. The second approach for fully drawn filaments is to conduct hot drawing on a separate machine. The low or medium orientated asspun filaments would first be wound onto bobbins, which are then transported to a separate drawing machine.

The structure–property relation of the major commercial polymers will be discussed later when dealing with the properties of CF yarns. At this point, it is useful to consider in more detail the classification of the melt-spun CF yarns. From the above description of the melt-spinning process, there are four basic melt-spun CF yarns, which will be described next.

Where extrusion is done separately from orientational drawing, i.e. the yarn is first extruded and then transferred to a separate machine for hot-drawing, the extruded filaments possess little or no orientation, and such yarns are classified as LOY (low oriented yarns). Relatively low melt-spinning speeds are used for producing these yarns. For example, PET LOY yarns would be spun at a speed of around 1500 m/min. After hot drawing, the yarns are classified as FDY (fully drawn yarns). FDYs are mainly used in weaving for furnishing fabrics.

The drawing system can be incorporated into the spinning line. The yarn produced would then be drawn to and have the appropriate degree of orientation. These yarns are referred to as SDY (spin-drawn yarns). The output or production speed would be approximate to the product of the extrusion speed and the total draw ratio.

Where the godet rollers below the spinneret are run at a much higher speed than is the case for LOY yarns, i.e. within the range of 2500 to 4000 m/min, the draw-down will be sufficient to partially orientate or pre-orientate the polymers parallel to the filament axis. This type of yarn is referred to as POY (partially oriented yarn), and is mainly used in one of a range of subsequent processes known as texturing or texturising, POY can also be used in draw warping for the weaving and warp knitting of fabrics. Higher draw-down speeds of 6000–7000 mm/in may be used to obtain highly oriented filaments that are not subjected to further drawing. These yarns are termed HOY (highly oriented yarns) when produced at a speed of 4000 to 6000 m/min and FOY (fully oriented yarns) at speeds above 6000 m/ min. FOY are usually made for technical textile applications (McIntyre, 1998). Although FDY/SDY yarns are used in textile and fashion fabrics, the majority of yarns for such applications are POY.

8.1.2 Wet Spinning

Not all polymers can be melt-spun, as some will thermally degrade rather than melt. In these cases, the polymer may be dissolved into a solution of sufficient viscosity to permit extrusion through a spinneret. The viscous solution (termed the dope) is extruded into a bath containing a second solution (called the spin bath or coagulation bath), which precipitates the polymer by diffusion and coagulates the polymer chains into continuous solid filaments.

The dope is usually prepared by dissolving and stirring the polymer in a heated solvent. The amount of polymer addition should be sufficient to give the dope adequate viscosity for the continuous extrusion of liquid filaments. The dope is de-aired, removing air bubbles that may occur during the mixing stage, and filtered to remove any impurities or partially dissolved polymer. The wet-spinning process is illustrated in Figure 8.3. The dope is forced by a metring pump through the holes of a spinneret immersed in the coagulating solution. The filaments are drawn by the first set of godets as they coagulate, initially into a gel state and then into a more solid state when the solvent diffuses into the solution.

The partially drawn or as-spun filaments are washed as they are driven around by the godets. On leaving the first set of godets, the filaments pass through a heated bath, where they are further attenuated in their solid state to straighten and align (i.e. orientate) the polymer chains with the fibre axis.

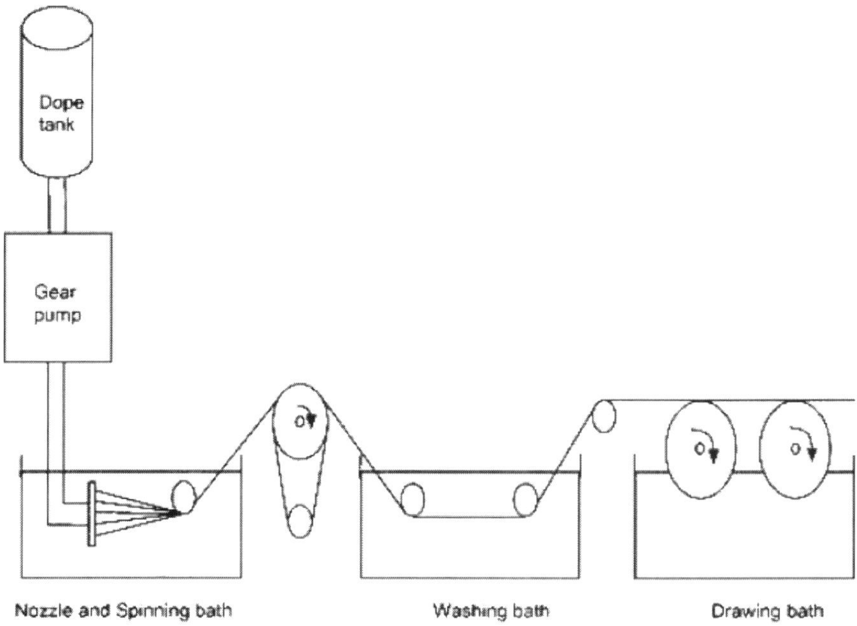

Figure 8.3 Schematic illustration of wet spinning (**Source:** http://www.tikp.co.uk)

They are then dried prior to winding into a package. Of the various textile polymers, viscose rayon is one example of a material converted into a CF yarn by the wet-spinning method. Figure 8.4 illustrates the full production process. The viscose dope – cellulose xanthate – is pumped by a 'metering' gear pump at 2.6 to 5 times atmospheric pressure. This is a much smaller pressure than in melt spinning, because the viscosity of the dope is significantly lower. Similar to melt spinning, the pump must deliver the dope to the spinneret at a constant rate to obtain a consistent filament diameter. The spinneret holes are usually 50–100 μm in diameter. The spinneret is submerged in an acidic coagulation/spin bath of sulphuric acid, sodium sulphate, zinc sulphate and water. The acid neutralises the cellulose xanthate by diffusion and precipitates the cellulose polymers, coagulating them into filaments.

To prevent corrosion by the acids used in the process, the spinnerets may be made of nonreactive metals such as platinum, tantalum or gold. The spin bath is made from lead sheets. The temperature and chemical concentration of the bath in relation to the dope are important factors in the actions of diffusion, polymer precipitation and coagulation. On leaving the spin bath, the filaments are passed through a drawing (stretching) zone in which stretching may range from 20% to 200%, depending on the spinning conditions, and are then washed, bleached and rewashed in a finishing stage to remove acid, salts and occluded sulphur. The spun yarns can then be dried and prepared for winding into a package.

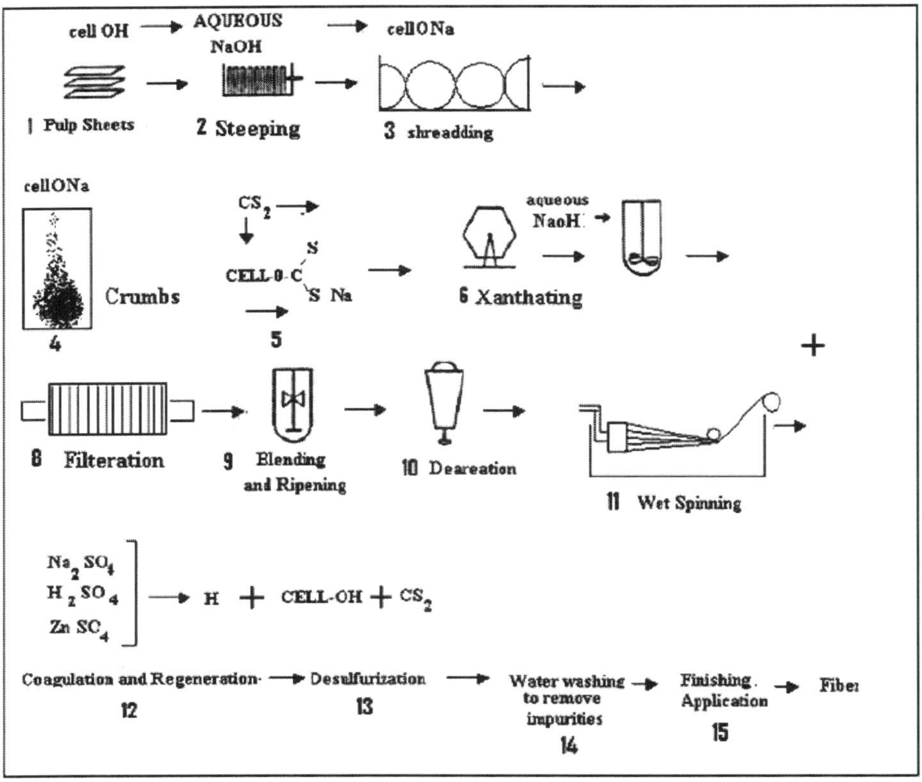

Figure 8.4 Schematic illustration of Wet spinning of viscose rayon (Source: http://textilelibrary.weebly.com)

8.1.3 Dry Spinning

A volatile solvent may be used to dissolve non-meltable polymers. Dry spinning is a process for producing CF yarns by dissolving a polymer in a volatile solvent to make a viscous solution that is extruded into a heated atmosphere to coagulate the polymer through solvent evaporation while stretching the filaments by means of drawing.

The polymers cellulose acetate and polyurethane (spandex) 4 are converted into CF yarns by the dry spinning method. The basics of the process are similar to that illustrated in Figure 8.5.

After the polymer is dissolved and the dope (which, again, is of much lower viscosity than with melt spinning) is filtered, de-aired and pre-heated, it is pumped through further filters and through the spinneret into a heated cabinet and gas flow. This may be air or an inert gas. Air is used for acetate. For polyurethane, nitrogen (N2) plus a solvent gas cause the polymer to react chemically to form solid strands. As the viscous filament streams enter the gas flow, there is an instantaneous evaporation of solvent from the surface of each stream, which initially forms a solid skin. The filaments then solidify with further evaporation as they pass through the gas flow. Simultaneous stretching by down-drawing can be applied as required, similar to melt spinning, so enabling orientation of the polymer chains along the fibre axis. At the bottom of the heated cabinet, the filaments converge into CF yarns and are wound onto packages by a ring and traveller system, which inserts a small degree of twist to hold the filaments together.

Usually the evaporated solvent is recovered to meet the requirements of safety, environmental regulation and economical processing.

8.2 Yarn Texturing

As obtained from the extrusion spinning processes, CF yarns are often described as flat yarns, meaning that they are inherently smooth and have no significant bulk or loft. With the exception of elastomers, they have little stretch at low applied forces and are lustrous unless altered by additives, as will be explained later. In this form they have a limited range of applications.

Therefore CF yarns are generally modified to make them more opaque and to add bulk and stretch at low tension, thereby imparting such fabric aesthetics as increased thickness, cover, softness, warmth, stretch and water vapour permeability (i.e. breathability). Usually modification is performed only on multifilament CF yarns. The filaments are subjected to a major change in their physical form by being crimped, coiled or looped along their lengths (see Figure 8.5).

The processes used to make the change in physical form are collectively called texturing. Various texturing methods are practised, most of which are only suitable for thermoplastic CF yarns as they involve heating the filaments to achieve the texture profile and cooling to retain it. The methods commonly referred to are the following.

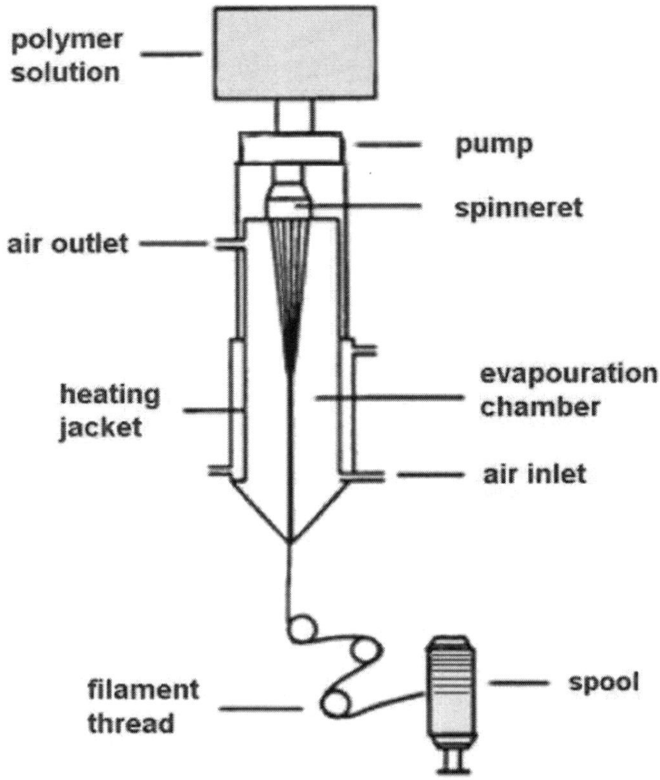

Figure 8.5 Schematic illustration of dry spinning (Source: http://www.tikp.co.uk)

- **Knife edge:** The filament yarn is heated and pulled across a blade at an acute angle. When the yarn is cooled and released, the retained internal stresses in the filaments cause them to collectively adopt a spring shape or curled ribbon appearance, i.e. the profile is heat-set.
- **Stuffer box:** The filaments pass through a heated box at a faster rate than they are removed (known as overfeed). This forces them to adopt a random wavy, crimped pattern while heated. The textured form is set by subsequent cooling.
- **Air jet:** High-speed overfeeding is also employed, but instead of using heat to effect the texture profile, compressed air is blown into the chamber, which causes the loose lengths of filaments in the yarn to separate and form entangled random loops. The entanglement retains the texture of random loops.
- **False twist:** The CF yarns are twisted and heated simultaneously and then untwisted when cooled, thus loosely retaining the heat-set helical shape of the twist.
- **Knit-deknit:** Filament yarns are knitted into the shape of a small-diameter tube and heat-set (i.e. heated and then cooled). The yarns are then de-knitted, giving them a wavy configuration.

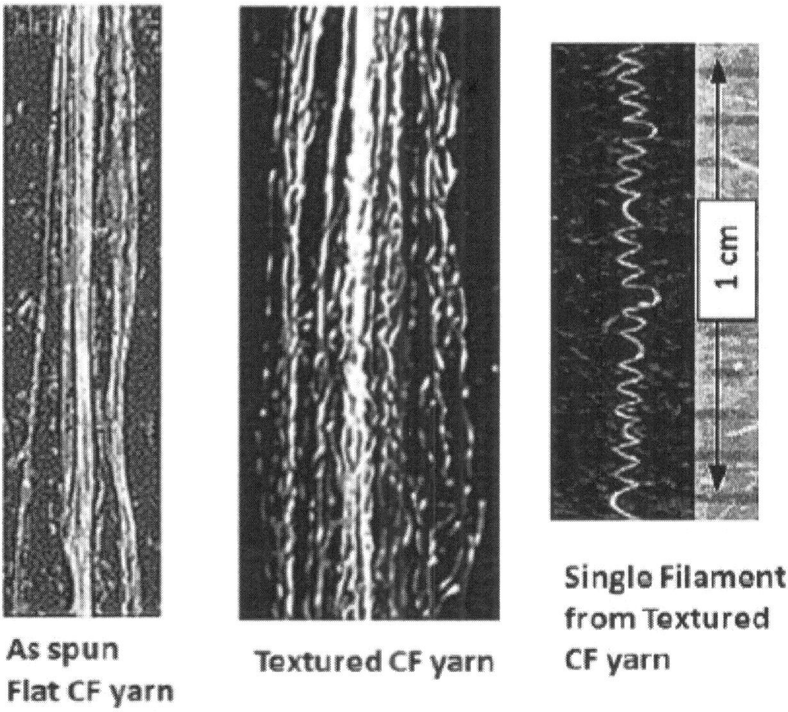

Figure 8.6 **Flat and Textured CF yarns** (**Source:** http://www.tikp.co.uk)

8.3 Bulk continuous-filament (BCF) technology

This is essentially an integrated process of melt spinning and texturing, using either a stuffer box or a hot-fluid jet. The filaments leaving the extruder is tightly bunched up into the heated stuffer box with hot compressed air or driven by the hot compressed air through an air jet at high speed and a high plasticising temperature. The yarn becomes irregularly kinked and is then cooled and set.

BCF yarns usually are comprised of coarse individual filaments of 8–30 dtex (7–27 denier) for nylon (PA and PA 6.6) or polypropylene (PP), i.e. PABCF and PPBCF yarns. Their main areas of applications are carpets and upholstery. Finer filaments of 3.3–5.6 dtex (3–5 denier) in nylon (PA and PA 6.6) are possible for producing PA-BCF yarns of around 600 dtex, which may be suitable for furnishings and outerwear. PET has a more complex heat transfer to the core of the filaments, but development work has enabled PET-BCF yarns to become available with significant growth in the US market. Since BCF technology is basically an integration of processes earlier described, it will not be considered further. From these brief descriptions of the above methods, texturing may be defined as a process by which continuous multifilament yarns are converted from their flat state to a greater bulked form, either by crimping and heat setting, twisting and heat setting, or the entanglement of loops. Although these methods are all in use, the most widely used is false-twist texturing followed by air-jet texturing.

8.3.1 False-Twist Texturing

False twist is the action by which twist is inserted into a yarn at the position of contact with a twisting device, and is then removed by the same device as the yarn leaves contact with it. A more detailed understanding can be acquired by comparing the actions involved in real and false twisting.

8.3.1.1 Real twist

When twist is inserted into a yarn, starting at its free end, the filaments adopt a helical shape or spiral, which may be clockwise or counter-clockwise in accordance with the direction of rotation of the twisting device. In a clockwise spiral, the inclination of the helix mirrors the inclination in the letter Z, and is therefore called Z-twist. In a counter-clockwise spiral, the twist is referred to as S-twist (see Figure 8.7). After the action of twisting, the inserted twist remains in the yarn.

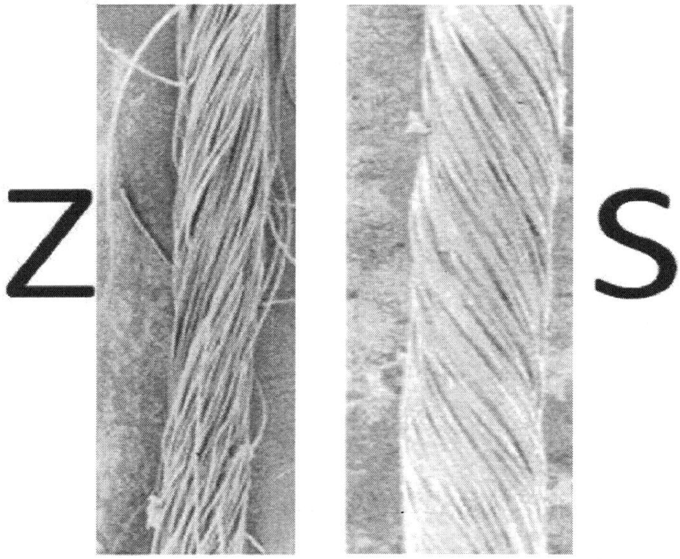

Figure 8.7 Photo-micrographs of Z and S twist yarns (Source: http://www.tikp.co.uk**)**

8.3.1.2 False twist

If the twisting device through which the yarn passes is located so that the Z-twisting action (the clockwise torque) occurs as the yarn runs into the device, and the S-twisting action (the counterclockwise torque) occurs as it runs out, then no twist will remain in the yarn. This is because the Z-twist will be removed by the counter-clockwise torque. This means that within a short time of starting the twisting process, Z-twist will be seen in the yarn as it runs into the device, but no twist will be present as it leaves. This is the false-twisting action, as illustrated in Figure 8.8.

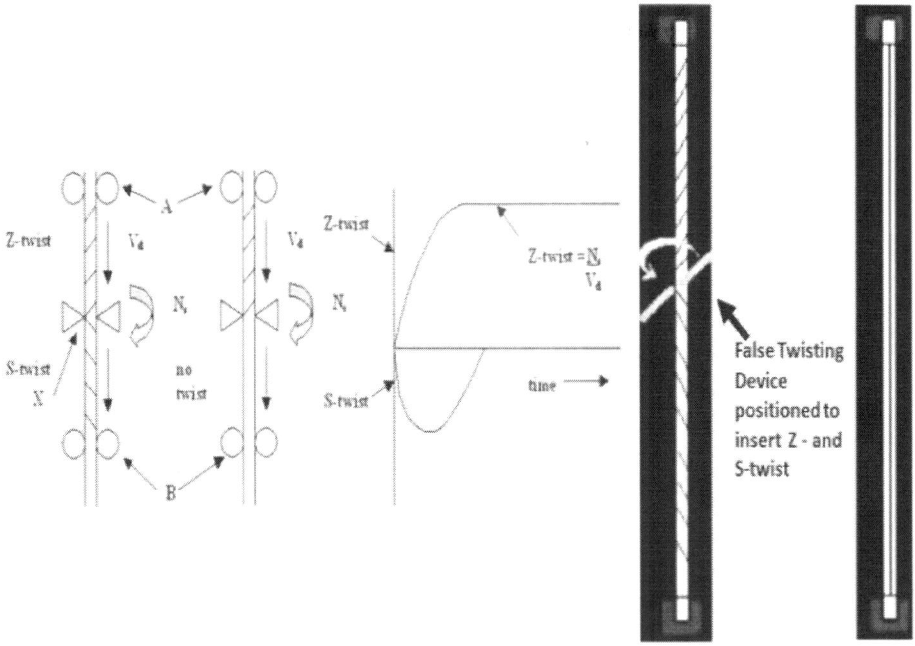

Figure 8.8 Schematic illustration of the false-twisting action (Source: http://www.tikp.co.uk)

The figure depicts the situation in which a CF yarn, nipped by two pairs of rollers at positions A and B, is driven at a linear speed of Vd m/min. The yarn is constantly twisted at the location X as it moves through the distance AB. If the twisting device is rotating at Ns rpm in the direction shown, then when viewed from A, along the length AX, it will appear to be turning clockwise.

However, viewed from B, along the length BX, it will appear to be turning counter-clockwise. Therefore, the same device will at any given time effectively twist the yarn length present within the zone AX in a clockwise direction, thus inserting a Z-twist whilst simultaneously twisting counter-clockwise the yarn length within the zone XB, inserting an S-twist. The Ztwist in the yarn length present in AX will rapidly increase to a constant value, equal to Vd/Ns turns per metre. In zone XB, the S-twist will increase to a maximum value and then decrease to zero, because each length of yarn moving from zone AX into zone XB will become untwisted by the counter clockwise torque present in zone XB.

If the yarn were to be heated above the Tg of the polymer whilst being Z-twisted in zone AX (Figure 8.9), and then cooled before being untwisted in zone XB, the spiral shape of the individual filaments would be retained, but they would become free of the twist compaction, resulting in a false-twist textured yarn. This would be bulky with considerable stretch at low applied force, depending on the level of false twisting.

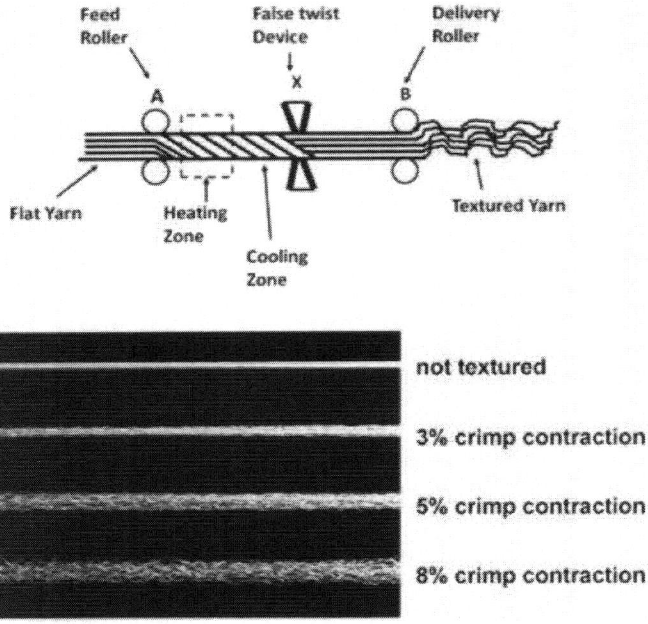

Figure 8.9 Illustration of false-twist texturing principle (Source: http://www.tikp.co.uk)

During the earlier years of false-twist texturing, FDY/HOY yarns were texturised using a rotating pin as the false-twist device. Figure 8.10 illustrates the pin twister as a hollow spindle with a central pin positioned across the tubular interior. The flat CF yarn passes down the tube and around the pin. The pin is rotated by a driven disc assembly.

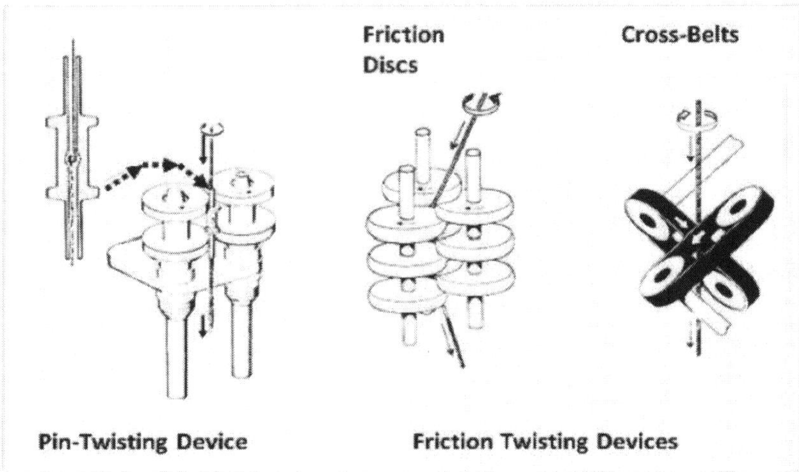

Fig.8.10 Schematic Illustrations of friction-twisting devices (Source: http://www.tikp.co.uk)

Partly oriented yarn may be utilised by the addition of a hot-draw zone before the heating/twisting/cooling zone, thus giving better process economics and product variety in terms of bulk and stretch (depending on the pre-drawing). The resulting yarn is then referred to as draw-textured yarn, or DTY. The pin-twister was later replaced by a friction-twisting device. As shown in Figure 8.11, an assembly of overlapping discs enables the CF yarn to be directly twisted by utilising the frictional contact between the yarn and the disc surface, thus replacing the need for a hollow spindle. Another type of friction-twisting device is based on driven cross-belts. These devices enable higher twisting speeds: 20 × 106 rpm as opposed to 8 × 105 for pin twisting. Depending on the polymer and yarn count, friction twisting enables production speed of up to 12,000 m/min to be attained.

DTYs are considered to be highly stretchable. In addition to the intrinsic extension of the constituent filaments, there is added extension attributable to the crimp imparted by the texturing process. Highly stretchable yarns can be made by applying high false-twist levels that give crimp extensions of 150–300%. However, where DTYs are required to have relatively low crimp extension while retaining a high bulk, an additional heating stage is incorporated after the roller-pair B (Figure 8.11). A third pair of rollers would then be used as output rollers.

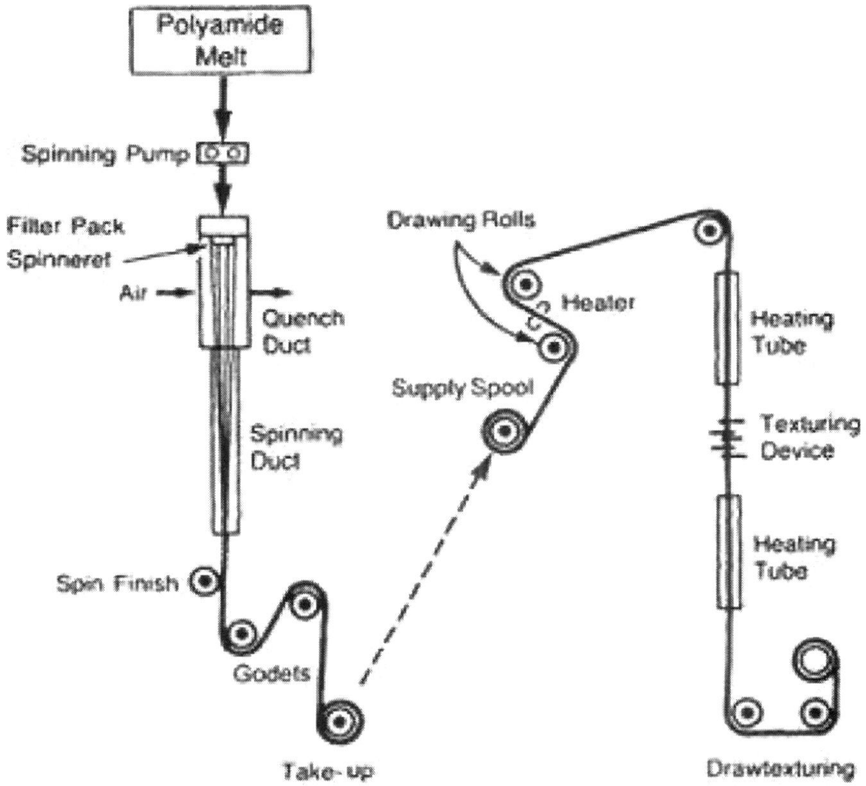

Figure 8.11 Illustration of the set-draw-textured yarn process

This heating stage gives an additional heat-setting treatment that reduces the intrinsic elasticity and crimp extension of the DTY. The yarn is then described as stabilised or set-DTY. An example of the set-DTY process is illustrated in Figure 8.11. This shows the melt spinning of nylon POY, which is then transferred to a draw texturing process. Here, the yarn is first drawn at a draw ratio of the order of 2.3–3.5 and a temperature of at least 50 °C (T_g = 40 °C). It is then subjected simultaneously to false twisting (friction disc device), heating and further drawing at a draw ratio of 1.1–1.5.

After false twisting, the DTY passes through the additional heating zone to become a set-DTY. The process is particularly useful for the manufacture of nylon textured yarn with a linear density in the range of 10 dtex/9 denier to 50 dtex/45 denier.

Thermoplastic multifilament yarns made from polyesters, nylons and polypropylene can usually be converted into DTYs. Polyester yarns are normally within the count range of 55.6 to 333 dtex (50–300 denier), with the emphasis on 83.3 to 166.7 dtex (75–150 denier). Nylon yarns are in the range of 16.7–122.2 dtex (15–110 denier); the majority of fine hosiery yarns are 24.4 dtex (22 denier) and coarser yarns are 77.8 dtex (70 denier). Polypropylene yarns are typically 7.7 to 100 tex (70–900 denier). Although false-twist texturing gives CF yarns a more natural feel, bulk and stretch, their uniform geometry lacks the natural appearance of staplespun yarns. Various techniques have therefore been developed to modify their appearance. An additional drawing zone using a hot pin may be placed before the texturing stage to impart irregular and repetitive changes to the filaments so that after texturing, the yarns display light and dark sections when dyed. If the draw ratio is cyclically varied from high to low during the main drawing stage and the twist insertion similarly varied, the textured yarn will have thick and thin places along its length. Although this property may be partially removed during downstream processing, a sufficient amount of the property is retained, say, during weaving to give the fabric a linen-like appearance.

8.3.2 Air-Jet Texturing

Air-jet texturing has become increasingly important because of its capacity to process CF yarns of any polymer type. For example, cellulosic rayon, which is not a thermoplastic, can be airjet texturised. A typical air-jet textured polyester/viscose yarn blend would comprise a FOY polyester of 80/75 Tex (72/75 denier) and a viscose CF of 80/24 dtex (72/24 denier).

Air-jet texturing is also capable of providing products with aesthetic characteristics (tactile handle) superior to most other texturing processes. It can be used to process FDY/FOY (to give air-textured yarn, or ATY) or with a pre-drawing stage in an integrated draw-texturing system to convert POY to a drawn-air textured yarn (DATY).

Various air-jet designs may be used to make ATY and DATY. Figure 8.12 depicts one of the widely used jets, referred to as an axial-jet (or venturi jet). The filament lengths to be texturised are fed into the jet without tension. This is achieved by ensuring that the speed at which they are fed into the jet is faster than the speed at which they leave (overfeed >1). The compressed air entering the jet then acts on the loose filament lengths. The compressed air is accelerated towards the outlet and pulls the filaments along. A spherical barrier is positioned close to the outlet, which causes the compressed air to become turbulent as it exits. The filament lengths in flow become separated and form a profusion of loops of various sizes, which entangle as the filaments are pulled together by the removal rollers to deliver the air- jet textured CF yarn.

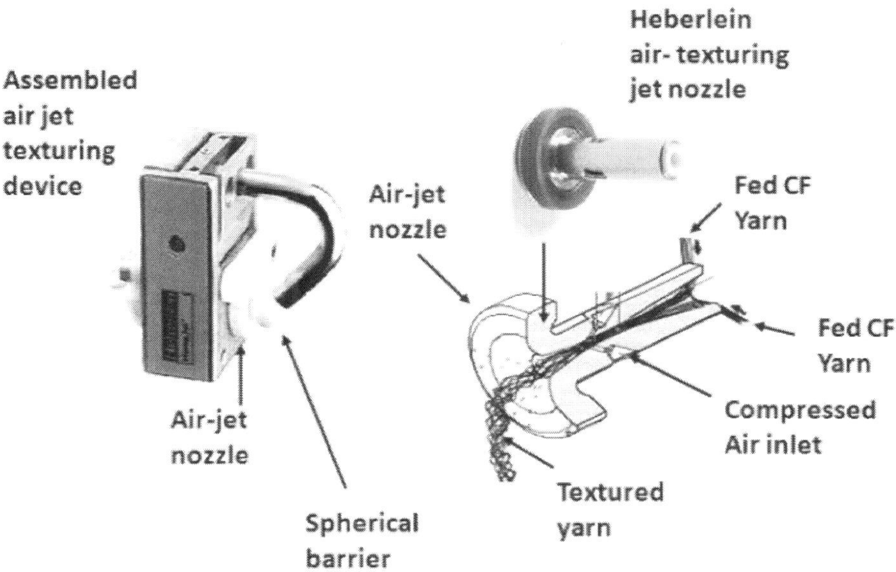

Figure 8.12 A Schematic illustration of Air-jet texturing device

A vortex-jet design may also be used. In this device, the compressed air has a spiraling turbulent flow to the outlet in which loose filament lengths are separated and form differentsized entangled loops as they exit the jet to become the ATY. These jets usually operate at pressures of about 10 bar (145 psi), and the actual texturing occurs directly at the exit of the jet within the turbulent air stream. To assist in the formation of loops, the filaments may be wetted prior to entering the jet, either by spraying or immersion in a water bath. This aids separation of the filaments and acts as a lubricant inside the jet.

POY is preferred for air-jet textured apparel yarns, largely because it is less expensive than FOY. The operating stages of a DATY machine would therefore be of a similar configuration to that of a DTY machine, but with the false-twisting device replaced by an air-jet system, as illustrated in Figure 8.12.

Two types of yarn can be produced with air-jet texturing: parallel yarns and core-effect yarns. The production of parallel yarns involves one or more POY being fed into the air-jet device with precisely the same overfeed for all the yarns used (Figure 8.14). Usually, an overfeed of 18–30% may be applied, depending on the end-use requirements. Parallel yarns are commonly used for apparel and cut pile plush fabrics.

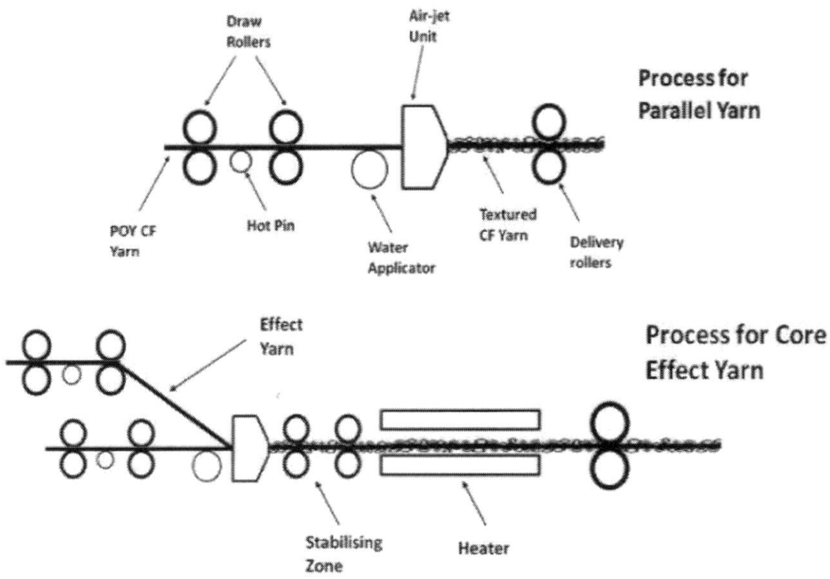

Figure 8.13 Schematic illustrations of Processes for parallel and core-effect air-textured yarn.

Core-effect yarns (or fancy yarns) are produced with two main components: a core yarn and an effect yarn. Both may consist of one or more POYs. The overfeed used for the core yarn(s) is always lower than that for the effect yarn(s): the former is normally between 5% and 15% whereas the latter can be up to 400%, depending on the end-use. For example, overfeeds of 8% core and 40% effect would be used for an apparel end-use such as nylon sportswear; while 20% effect yarn overfeed would be appropriate for upholstery fabrics for domestic use and car seats.

Figure 8.13 shows that the drawing zones for both yarn types have a heating element, which may be a heated pin, godet or plate. The heating element is essential for polyester POY, but nylon and polypropylene POY may be cold drawn. Note that in the case of core-effect type yarns, the two yarn paths must have their own drawing zone and heating element.

Although CF yarns of virtually any polymer can be air-jet textured, the number of filaments and the dtex/denier per filament are very important; a high number of filaments of low dtex/denier per filament is preferred. Yarns with a higher than 3.3 dtex (3 denier) per filament are rarely air-jet textured. Currently, the practise is to use between 1.1 and 2.2 dtex (1 and 2 denier) per filament (dpf), the most popular yarns tending towards 1 dpf or finer.

Air consumption is a significant cost factor in the process and is kept as low as possible. Therefore a variety of air-texturing jets are available to cover the different ranges of dpf. Vortex-type jets are generally used for products that require less than 100% overfeed, whereas the venturi type of jet may be used with overfeeds of up to several hundred percent. Loop size and stability are important factors, especially in finer yarns. Smaller loops are more stable and give a bulkier feel than larger loops. As shown in Figure 8.13, the textured yarn can be heat-treated to increase the stability and shrink the loops as they leave the air jet. The temperatures must be sufficient to ensure that the constituent filaments are adequately treated. For finer yarns, sufficient setting temperatures are usually 230–240 °C. If required, an additional drawing zone placed between the air jet and the heater may also be used, ensuring small-sized loops are obtained.

A further addition to the process line can be made that converts small loops into hairs, thus simulating the structure of staple-spun yarns, which have short protrusions of fibre ends from the yarn surface.

The process used for achieving this staple fibre-like yarn is called Texspun and has been developed by the machine manufacturer Barmag (Air Jet Texturing and Fabrics). The Texspun device is located in the heat-stabilising zone. As the yarn passes through the device, the loops are torn and free fibre ends, similar to staple fibres, are produced.

8.4 Bulk Continuous Fibre (BCF) Technology

8.4.1 Twisting/Plying of Continuous-Filament Yarns

A single multifilament FDY yarn (flat yarn) can be twisted so that the twist holds the filaments together and imparts some degree of stretch as a result of the twist helix in combination with the intrinsic stretch of the spun polymeric material. When a small amount of twist (40 tpm or less) is used to hold together the filaments of a multifilament FDY yarn, it is referred to as producer twist, since it is usually inserted at the yarn-production stage. The more common practice is to intermingle the filaments with a special air jet.

Twist levels can range from that of producer twist to over 1000 tpm, and the higher the value, the more monofilament-like the multifilament yarn will become. Two or more yarns (mono- or multifilaments or combinations) may also be twisted together to produce greater bulk, or in the case of textured yarns, to alter the stretch and mechanical performance.

When two or more yarns are twisted together, the action is called doubling, folding, plying, or ply twisting. The resulting yarn is termed a doubled, folded or plied yarn (e.g. 2-ply, 8-ply). For example, PET yarns as fine as 49 dtex and up to 110 tex are often labelled as PET 49 dtex 1 × 2 (i.e. 2 yarns plied) and PET 1100 dtex 1 × 3 (i.e. 3 yarns plied) (Trougott Baumann). In plied yarns, each yarn may be individually Z-twisted, and then twisted together in the Sdirection, thus alleviating a tendency for the plied yarn to snarl. Plied CF yarns are used in embroidery and for sewing threads.

There are three basic methods used in filament yarn twisting: up-twisting, down twisting and two for- one twisting. Figure 8.14 shows the principal features of each method.

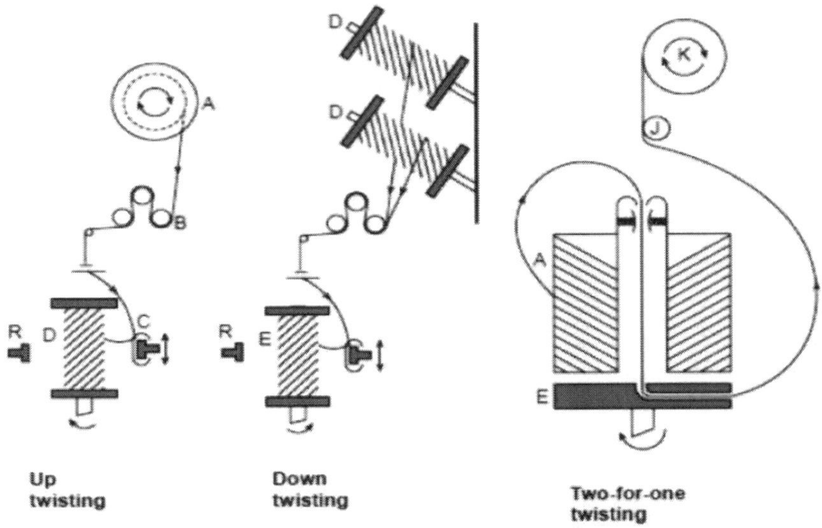

Figure 8.14 Schematic illustration of Yarn-twisting processes

In up-twisting and down-twisting, a ring (R) and traveller (C) arrangement may be used. The traveller is the 'C' metal or plastic clip that is loosely clipped onto the ring profile. The traveler circles the ring, and effectively each circulation will insert one turn of twist into a single yarn or a combination of two or more yarns.

In up-twisting, the circulation of the traveller around the ring is obtained by placing a bobbin (D) of CF yarn on a spindle (S) concentric with the ring, threading the yarn from the bobbin through the gap between traveller and ring and via a set of delivery rollers (B) onto a take-up bobbin (A). The rollers remove yarn from (D) and deliver it to (A). As the spindle rotates in the direction indicated, the yarn is easily removed and the take-off rollers simultaneously pull the yarn through the traveller/ring gap, causing the traveller to be dragged around the ring. The circulation of the traveller causes the length of yarn between traveller and take-off rollers to move radially outwards. This is termed ballooning, because the speed of the circulating length gives the optical illusion of an inflated balloon. For each rotation of the traveller, one turn of twist is inserted in the length. The delivery rollers then feed forward the twisted yarn for it to be wound onto the take-up bobbin by a drive system, the process being continuous until all the yarn is removed from the bobbin (D).

Up-twisting is suited to single CF yarn, but if more than one yarn is to be twisted, they may be initially wound together onto the up-twisting supply bobbin (D). This is referred to as assembly winding. In down-twisting, two or more yarns on separate supply bobbins are pulled by the delivery rollers, which operate in the opposite way to the up-twisting action. The yarns pass via the ring and traveller onto a bobbin mounted on the spindle. The spindle rotates in the opposite direction to that of up twisting. The yarn is now wound

onto the spindle bobbin, instead of being unwound from it. As before, the yarn drags the traveller around the ring as it is being wound onto the bobbin, and the twist is inserted within the yarn length between the traveller and the delivery rollers. The winding action occurs because the friction between the ring and traveller causes the traveller to always move slightly more slowly than the surface speed of the rotating bobbin.

Figure 8.14 also illustrates the principle of two-for-one twisting. The yarn or yarns are withdrawn from a stationary supply package or packages (A), passed through the centre of the package(s) and through the centre of a rotating mechanism (E), driven by the spindle (S). The yarn(s) then balloons around the supply package(s) to be delivered by a pair of rollers (D) and wound onto the take-up package (K). The yarn (or yarns) is twisted once as it passes down the package centre (A), and a further turn is inserted between E and the delivery rollers (D), resulting in two turns of twist for each turn of the spindle (E).

Wrapping one yarn around another to produce a covered yarn may be considered as a special case. This process is normally used to produce elasticated filament yarns, which are PU-CF yarns covered with a textured CF yarn or staple-fibre yarn7, making them easier to weave or knit and providing them with better tactile characteristics and fabric drape. Such yarns are used in woven fabrics, narrow fabrics, flat knitting and circular knitting, and can be found in gloves, sweaters, intimate apparel, swimwear, hosiery and socks (BeYarns). With wrapping, snarling should not occur. Therefore wrapping with a single-yarn covering can be carried out with either a Z- or an S-twist. It is common to use a Z-twist for elasticated fabrics. Double covering may be employed where the cover yarns are wrapped in the opposite directions, one being the outer covering (S-twist), and the other the inner covering (Z-twist). The different twist directions improve the tactile properties. Double covering may be used for socks and underwear to enable a more accurate body profiling.

8.4.2 Metallised Yarns

A less conventional method of making filament yarns is to first extrude (using long narrowgap dies instead of spinnerets) a thermoplastic polymeric material into a thin, wide sheet or film similar to that employed for packaging. Metering pumps are generally not used. The film is normally quenched in a water bath, or on chilled rolls. On leaving the bath, the film is blowdried. It is then cut into narrow ribbons, drawn and annealed. Annealing is where the film is re-heated for a short period of time below the glass transition temperature of the polymer, in order to relieve internal stresses and allow the polymer chains to partially relax. The film is then gradually cooled to room temperature. The oriented polymer chains readily allow slits to be made along the length of the ribbons, i.e. the ribbons become fibrillated. These can then be twisted to make split film (or fibrillated) yarns. Although the majority of these yarns are produced for ropes and other technical textile applications, the process can be adapted for metallised yarns that are used for their visual aesthetics. Typically, a PET film is metallised by depositing a thin layer of aluminium or gold, and then lacquered with a protective epoxy resin that maintains the shiny look of the metallised surface. If a coloured shiny look is required, the requisite dye is added to the resin prior to coating. The film can then be cut into ribbons that are further reduced to strips of around 0.2–0.3 mm width (M-Type Metallised Yarns) and can be plied with other CF yarns or staple-fibre yarns for decorative end-uses.

References and Bibliography

1. Elmogahzy, Y., and Chewning, C., Jr, *Fiber To Yarn Manufacturing Technology*, Cotton Incorporated Publication, Cary, NC (cottoninc.com), 2001.
2. Leifeld, F., Modern spinning preparation. Part A: Blow room and carding machines, systematically correlated, *German Technology Symposium & Exhibition*, Bangkok, Thailand, November 1993.
3. Lünenschloss, J., and Brockmanns, K. J. Mechanisms of OE-friction spinning, *Int. Textile Bull.*, 3, 29–59, 1985.
4. Fehrer, E., The present state-of-the-art in friction spinning, *American Textile International*, 86–89, December, 1986.
5. Stalder, H. and Soliman, H. A., A study of the yarn formation process during friction spinning, *Melliand*, No. 2, E44–E46, 1989.
6. Brockmanns, K. J. and Johnson, N., The influence of fiber feeding arrangements on the properties of friction spun yarn, *Int. Textile Bull.*, 3, 27–34, 1986.
7. Brockmanns, K. J., Possibilities of OE-friction spinning, *Textile Month*, 10–16, December 1984.
8. Lord P.R., 2003, *Handbook of yarn production*, Woodhead Publishing, Cambridge.
9. Merino, 2008, Environmental Assurance Program (EAP), The Merino Co., Melbourne, Australia.
10. Mizerovskii L.N., 2000, Improved spin finish for synthetic fibres: a new approach to solving the problem, *Fibre Chemistry*, 32(3), 221–260.
11. Oldrich P., Reichstadter B., 1979, *Processing of Polyester Fibres*, Elsevier, Amsterdam.

Index

G

H

I

K

L

M

N

P

Q

R